# Lecture Notes in Nanoscale Science and Technology

Volume 25



More information about this series at http://www.springer.com/series/7544

Nebojša Nešković • Srdjan Petrović
Marko Ćosić

# Rainbows in Channeling of Charged Particles in Crystals and Nanotubes

Springer

Nebojša Nešković
Laboratory of Physics
Vinča Institute of Nuclear Sciences
University of Belgrade
Belgrade, Serbia

Srdjan Petrović
Laboratory of Physics
Vinča Institute of Nuclear Sciences
University of Belgrade
Belgrade, Serbia

Marko Ćosić
Laboratory of Physics
Vinča Institute of Nuclear Sciences
University of Belgrade
Belgrade, Serbia

ISSN 2195-2159 ISSN 2195-2167 (electronic)
Lecture Notes in Nanoscale Science and Technology
ISBN 978-3-319-61523-3 ISBN 978-3-319-61524-0 (eBook)
DOI 10.1007/978-3-319-61524-0

Library of Congress Control Number: 2017944204



Printed on acid-free paper

This Springer imprint is published by Springer Nature
The registered company is Springer International Publishing AG
The registered company address is: Gewerbestrasse 11, 6330 Cham, Switzerland

# Prologue

This book is devoted to the rainbows appearing in channeling of charged particles in crystals and nanotubes and to their various applications. We begin with a brief report on the optical rainbow effect, occurring in scattering of light from water droplets and similar targets. It is followed by a short review of the rainbows occurring in scattering of atoms by atoms or molecules, nuclei by nuclei, and atoms by crystal surfaces.

Then, we give a brief historical presentation of the process of ion channeling in crystals, ending with a very concise description of the analytical theory of axial ion channeling. This theory was developed by the methods of statistical mechanics employing the continuum string model and the assumption of statistical equilibrium in the transverse plane. The interaction potentials used to treat ion channeling are also introduced. This is followed by a condensed description of the ways of computer simulation of the process – using the continuum string model or the binary collision model. The former model has been used in the majority of studies presented in the book.

Upon that, we concentrate on the crystal rainbows, which were discovered in a theoretical analysis of ion transmission through an axial channel of a very thin crystal. The approach was via the ion-molecule scattering theory. In the created model, called the model of crystal rainbows, the continuum string model was included implicitly, and the assumption of statistical equilibrium in the transverse plane was avoided. Analysis showed that the effect is a gestalt effect. The crystal rainbows were observed experimentally soon after the discovery. Those were the first quantitative measurements of ion channeling in thin crystals, leading to a precious scaling law that has been confirmed in numerous later studies. According to this law, the evolution of the spatial and angular distributions of channeled ions is divided into cycles determined by the change of the reduced crystal thickness, which have been named the rainbow cycles. This is followed by a description of the effect of zero-degree focusing of channeled ions, being an effect of angular focusing, which marks the ends of the rainbow cycles. Further, we introduce the theory of crystal rainbows, which was formulated as a generalization of the model of crystal rainbows, providing a proper treatment of ion channeling in axial channels of crystals that are not necessarily very thin. Besides, we demonstrate that the crystal rainbow effect can be accurately modeled using catastrophe theory, i.e., it is a catastrophic effect.

Next, the results of a thorough analysis of a sequence of high-resolution measurements of the angular distributions of protons of kinetic energies between 2.0 and 0.7 MeV channeled in a 55 nm-thick (100) Si crystal are presented. Those extraordinary measurements have proven to be crucial for verification of the theory of crystal rainbows as the proper theory of ion channeling in crystals. Within that study, a morphological method for producing very accurate ion-atom interaction potentials has been developed. In continuation, we explore the process that is inverse to the ion transmission process under consideration and, thus, tackle the question of its multiplicity, directly connected to the essence of the crystal rainbow effect, being the ion focusing along a line reflecting the symmetry of the illuminated crystal. The consideration includes modeling using catastrophe theory. Besides, we present a detailed study of the effect of superfocusing of channeled ions, being an effect of spatial focusing, which occurs in the middle of each rainbow cycle. It is explained how the effect can be used for subatomic microscopy. The effect has been recently observed experimentally – it has been demonstrated that the diameter of a proton beam in a Si crystal can go down to ~20 pm. Further, we prove that the doughnut effect in ion channeling, observed with tilted crystals, is in fact a rainbow effect. This is done using a sequence of high-resolution measurements of the angular distributions of transmitted protons.

It has been shown theoretically that rainbows can also occur in ion channeling in nanotubes. Here, we describe the rainbow effect in transmission of 1 GeV protons through a bundle of very short (10, 10) achiral single-wall carbon nanotubes. Then, the results of a study of transmission of protons of the same kinetic energy through a bent very short (11, 9) chiral single-wall carbon nanotube are presented. We show that, in principle, it is possible that passive elements composed of bundles of bent carbon nanotubes substitute dipole magnets within high-energy accelerator facilities. Upon that, we analyze in detail the spatial and angular rainbow effects in transmission of the same proton beam through straight very long (11, 9) carbon nanotubes. All three studies have been performed using the theory of crystal rainbows.

In addition, we present a quantum mechanical consideration of transmission of 1 MeV positrons through very short (11, 9) carbon nanotubes. The time-dependent Schrödinger equation is solved, and the spatial and angular distributions of transmitted positrons are calculated using a computer simulation method. The initial positron beam is represented as an ensemble of noninteracting Gaussian wave packets. We examine the principal and supernumerary primary rainbows appearing in the spatial and angular distributions. We also give a detailed explanation of the way of their generation, which includes the effects of wrinkling of each wave packet during its deflection from the nanotube wall and of its concentration just before a virtual barrier lying close to the corresponding classical rainbow. In addition, the wave packets wrinkle in a mutually coordinated way.

We would like to thank our colleague Milan Rajčević for helping us prepare the figures.

Authors,
March 2017

# Contents

# Chapter 1
# Rainbows

The optical rainbow effect appears in scattering of light in the limiting case of small wavelengths. When the scattering process involves sunlight and water droplets in the atmosphere, one talks about the meteorological rainbows. The history of man's exploration and apprehension of the phenomenon is very long and rich. Its first explanation emerged way back in the middle of the fourth century BC. However, a satisfactory quantitative theory of the effect did not appear before the end of the 1960s.

In the beginning of the 1950s, it was discovered that, in elastic scattering of atoms by atoms or molecules, an effect analogous to the meteorological rainbow effect could occur. The first measurement of the phenomenon, which was named the atomic rainbow, was conducted about 10 years later. After that, two similar effects were observed experimentally in elastic nuclear scattering, one in elastic scattering of atoms from crystal surfaces and one in transmission of ions through crystal channels. These phenomena are referred to as the Coulomb, nuclear, surface, and crystal rainbows, respectively. The rainbow effect has also been observed in inelastic and reactive collisions of atoms with molecules, inelastic nuclear collisions, collisions with transfer of nucleons, and inelastic collisions of molecules with surfaces. All these effects are designated as the particle rainbow effects. In this book, we shall consider very briefly the atomic, Coulomb, nuclear, and surface rainbows, and concentrate at length on the crystal rainbows. The atomic, Coulomb, and nuclear rainbow effects are similar to each other. This is a consequence of the fact that, in these cases, the interaction potential is rotationally symmetric. On the other hand, one can find similarities between the surface and crystal rainbows. In these cases, the interaction potentials, not being rotationally symmetric, resemble each other.

At the end of the 1980s, a unique meeting of a number of major researchers of the rainbow phenomena in various fields was held in Cavtat, Yugoslavia.



N. Nešković et al., *Rainbows in Channeling of Charged Particles in Crystals and Nanotubes*, Lecture Notes in Nanoscale Science and Technology 25,
DOI 10.1007/978-3-319-61524-0_1

Its proceedings enable one to compare all these effects with each other in order to see their similarities and differences, and learn about the mathematical apparatus necessary for their description [1].

## 1.1 Optical Rainbows

The meteorological rainbow phenomenon comprises more components: the primary, secondary, and supernumerary rainbows. The primary rainbow is a bright circular bow seen in the sky at an angle of about 42° relative to the horizon. This bow consists of six concentric bows of different colors – the innermost bow is purple, then come the blue, green yellow, and orange ones, and the outermost bow is red. The secondary rainbow is a bright circular bow occurring in the sky on the outer side of the primary rainbow at an angle of about 50° relative to the horizon. This bow is also composed of six concentric bows of different colors, but the order of colors is opposite to the one in case of the primary bow. The region between the primary and secondary rainbows is dark. The supernumerary rainbows are the bright circular bows appearing on the inner side of the primary rainbow. The secondary and supernumerary rainbows are considerably less intense than the primary rainbow, and, usually, it is hard to observe them.

It seems that the first explanation of the meteorological rainbow effect was given by Aristotle in the middle of the fourth century BC [2]. It was related to the primary rainbow only. He thought that the rainbow appeared as a consequence of the reflection of sunlight from a cloud as a whole, with the reflected light rays forming a cone. On the basis of that, connecting the positions of the observer and the reflected rays, Aristotle explained the circular shape of the rainbow. One of the most important contributions to the explanation of the effect came from Theodoric of Freiburg in the beginning of the fourteenth century [2]. He abandoned the Aristotle's idea about the rainbow appearing due to the reflection of sunlight from a cloud as a whole, and assumed that each water droplet within the cloud produced a rainbow. Theodoric proved that assumption in an experiment in which a spherical glass vessel filled with water, representing an enlarged droplet, was illuminated with light.

R. Descartes conducted an experiment similar to that of Theodoric and analyzed in detail the process of scattering of light from a spherical water droplet [2]. Figure 1.1 gives a scheme of the process of reflections and refractions of a light ray on the droplet for one value of its impact parameter, $b$ [2]. The incoming ray lies in the vertical plane dividing the droplet in two equal parts. On the droplet boundary surface, the incoming ray is partly reflected and partly transmitted with the directions of propagation of the reflected and transmitted rays determined by the laws of its reflection and refraction, respectively [2]. Descartes found that the rays going out from the droplet could be clearly classified on the basis of the number of their reflections inside the droplet. In Fig. 1.1, the outgoing rays corresponding to the direct reflection from the droplet, and to no reflection, one reflection, two

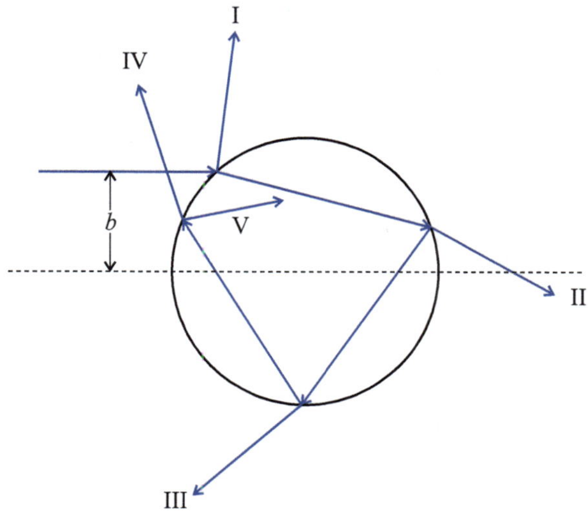


**Fig. 1.1** A scheme of the process of light scattering from a spherical water droplet. The incoming light ray, whose impact parameter is $b$, lies in the vertical plane dividing the droplet in two equal parts. The outgoing rays from classes I, II, III, IV, and V are presented

reflections, and more than two reflections inside the droplet are denoted by I, II, III, IV, and V, respectively.

Figure 1.2 gives the dependences of the scattering angle of the light ray from the water droplet, $\theta$, which is the angle between the directions of propagation of the incoming and outgoing rays, on $b/R_d$, where $R_d$ is the droplet radius, for the outgoing rays from classes III and IV that were obtained by Descartes [2]. It is evident that, in case of the outgoing ray from class III, the dependence has a minimum at $\theta_{r1}$=138°, and that in case of the ray from class IV, the dependence has a maximum at $\theta_{r2}$=130°. Since the distribution of impact parameters is uniform, relatively large intervals of $b$ are mapped to relatively small intervals of $\theta$ in the vicinities of these scattering angles. As a result, the scattered light accumulates at $\theta_{r1}$ and $\theta_{r2}$, and two bright circular lines appear on the screen placed between the droplet and light source. He realized that the corresponding angles at which those bright lines were seen equaled $\pi - \theta_{r1}$=42° and $\pi - \theta_{r2}$=50°, and concluded that the bright lines represented the primary and secondary rainbows, respectively. As expected, the secondary rainbow was less intense than the primary rainbow, which was clearly attributed to the additional reflection of the light rays inside the droplet in that case. It should be mentioned that $\theta_{r1}$ and $\theta_{r2}$ do not depend on $R_d$ [2]. The Descartes' results also showed that the outgoing rays from classes III and IV could not access the region of scattering angles between $\theta_{r1}$ and $\theta_{r2}$. This region has been named Alexander's dark band, after Alexander of Aphrodisias, who was the first to observe it, in the beginning of the third century. This means that the regions of $\theta$ above and below $\theta_{r1}$ are the bright and dark sides of the primary rainbow, and the regions of $\theta$ below and above $\theta_{r2}$ the bright and dark side of the secondary rainbow, respectively.

In analyzing the intensity of light scattered from the water droplet as a function of $\theta$, $I(\theta)$, Descartes found that it tended to infinity when $\theta$ changed toward $\theta_{r1}$ or $\theta_{r2}$ from the bright side of the rainbow. This is illustrated in Fig. 1.3 for the primary

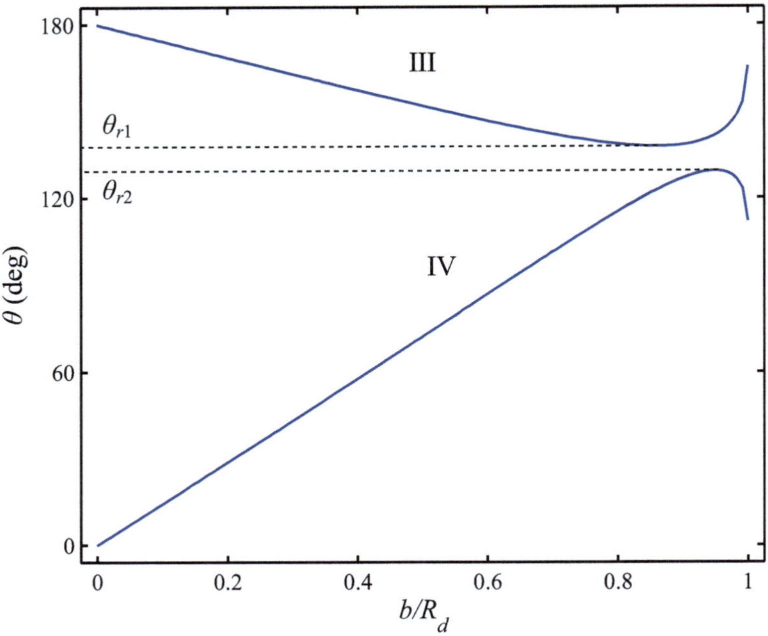


**Fig. 1.2** Dependences of the scattering angle of the light ray from the spherical water droplet on $b/R_d$ for the outgoing rays from classes III and IV. The former dependence has a minimum for $\theta_{r1}$=138° while the latter dependence has a maximum for $\theta_{r2}$=130°

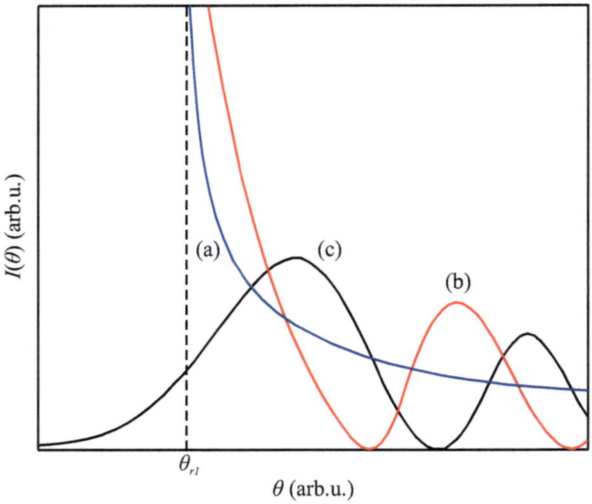


**Fig. 1.3** Intensities of light scattered from a spherical water droplet on $\theta$ in the vicinity of $\theta_{r1}$ obtained by (**a**) Descartes, (**b**) Young, and (**c**) Airy; the values of $I(\theta)$ are given in arbitrary units (Adapted from Ref. [2])

rainbow [2]. It has been established that the accumulation of the scattered light also happens after its three or more reflections inside the droplet [2]. These effects are referred to as the higher order meteorological rainbow effects, and they are very hardly observed.

The fact that the primary and secondary rainbows consist of the bows of different colors was explained by I. Newton [2]. He did that on the basis of the effect of dispersion of light, according to which the velocity of light in a medium depended on its characteristics. He calculated the primary and secondary rainbow angles for different colors and found that the obtained values agreed with the values that had come out from his experiments. With that contribution, Newton completed the explanation of the meteorological rainbow effect within geometrical optics.

It is clearly seen in Fig. 1.2 that, on the bright side of the primary rainbow, there are two values of $b$ giving one value of $\theta$. Thus, the effect of interference of the light scattered from the water droplet can occur, resulting in the alternate reinforcements and cancellations of the intensity of scattered light as a function of $\theta$ above $\theta_{r1}$. T. Young observed that fact in the beginning of the nineteenth century and used it to explain the meteorological supernumerary rainbow effect [2]. It should be emphasized that the scattering angles at which the supernumerary rainbows appear depend on $R_d$ [2]. The larger the droplet, the smaller is the difference between these angles. The supernumerary rainbows also exist associated to the secondary rainbow, for $\theta$ below $\theta_{r2}$, but it is very hard to observe them. Young's theory enabled the explanation of all the components of the meteorological rainbow effect, but not quantitatively. According to this, as in the case of Descartes' explanation, the intensity of scattered light at $\theta_{r1}$ and $\theta_{r2}$ was infinite. This is illustrated in Fig. 1.3 for the primary rainbow.

The quantitative theory of the meteorological rainbow effect was founded by R. Potter and G. B. Airy in the middle of the nineteenth century [2]. Potter analyzed in detail the accumulation of the light rays going out from the water droplet for $\theta = \theta_{r1}$ and found that $\theta_{r1}$ in fact defines the envelope of the outgoing rays from class III. Starting from those results, Airy formulated a theory of the effect based on the Huygens-Fresnel principle, given in the beginning of the nineteenth century [3]. The shape of his dependence of the intensity of scattered light on $\theta$ in the vicinity of $\theta_{r1}$ is shown in Fig. 1.3 [2]. One can see that, in this case, a maximum lying at the scattering angle somewhat larger than $\theta_{r1}$ appears. It corresponds to the primary rainbow. The second maximum of the dependence corresponds to the first associated supernumerary rainbow. In Alexander's dark band, i.e., for $\theta < \theta_{r1}$, the intensity of scattered light decreases quasiexponentially. One says that the maxima of the dependence show the diffractive character, and its quasiexponential decrease the refractive character of the scattering process under consideration. It is important to note that the positions and widths of the maxima depend on the ratio $\kappa_d = 2\pi R_d/\lambda$, where $\lambda$ is the wavelength of light, which is called the droplet size parameter and changes in practice from about 100 to several thousands [2]. Airy's theory gives satisfactory results for $\kappa_d > 5000$.

The conditions for formulation of an exact theory of the meteorological rainbow phenomenon were met in the middle of the nineteenth century, when Maxwell's

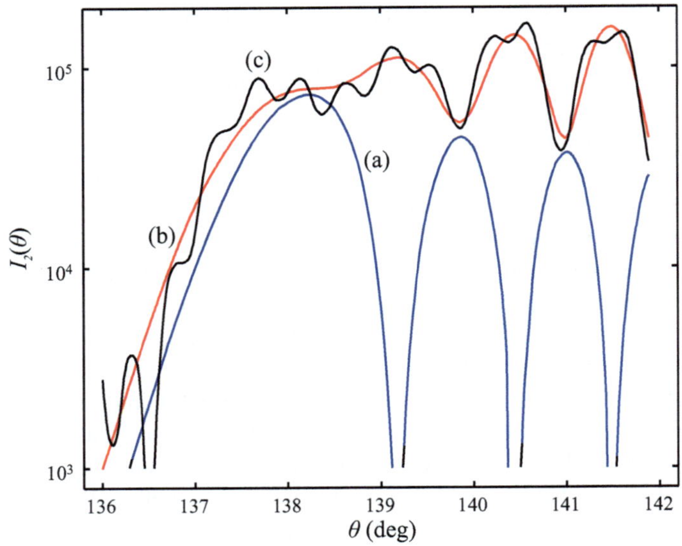


**Fig. 1.4** Intensities of light scattered from a spherical water droplet without the directly reflected component in the region of $\theta$ around $\theta_{r1}$ for the parallel polarized light and $\kappa_d$=1500 obtained using (**a**) Airy's theory, (**b**) Nussenzveig's theory, and (**c**) the exact theory (Adapted from Ref. [2])

theory of electromagnetic radiation emerged. The theory was created by G. Mie and P. J. W. Debye in the beginning of the twentieth century [2]. According to it, the scattering amplitude of light, whose square equals the intensity of light scattered from the water droplet, is an infinite series of the functions representing the so-called partial waves, each of which is determined by one value of the angular momentum of light. However, the series converges slowly – the number of terms that must be included in the calculation to attain a satisfactory accuracy is about $\kappa_d$. Also, the theory does not allow one to analyze separately the components of the effect. Therefore, usually, its use is not convenient.

A satisfactory quantitative theory of the meteorological rainbow effect was created by Nussenzveig [4–6]. Using Watson's transformation, he succeeded in reducing the problem of the infinite series appearing in the above-mentioned exact theory of the effect to the problem of an integral in the complex plane whose solution contains a small number of terms corresponding to the saddle points and poles of the integrand. In this theory, the angular momentum of light is a complex variable. It enables one to analyze separately the components of the effect. Figure 1.4 shows a comparison of the intensities of light scattered from the water droplet without the directly reflected component obtained using Airy's, Nussenzveig's, and exact theories in the region of $\theta$ around $\theta_{r1}$ for the parallel polarized light and $\kappa_d$=1500 [2, 7]. It is evident that the deviation of Airy's theory from the exact theory is considerable, especially in the region of supernumerary rainbows. On the other hand, the agreement of Nussenzveig's theory with the exact theory is satisfactory. The fluctuations of the dependence of the intensity of scattered light on $\theta$ generated by the exact theory come from the light rays from class V, which undergo more than two reflections inside the droplet. It is worth mentioning that, in the case of normally polarized light, the deviation of Airy's theory from the exact theory is smaller than that in the case of parallel polarization.

Besides Nussenzveig, Berry [8, 9] and Nye [10–12] have made exceptional contributions to the understanding of the optical rainbow effect. Berry investigated the possible shapes of the rainbows in the general case of scattering of light, in which the target was not spherical. He found that the obtained shapes could be classified using catastrophe theory [13] as a general theory of models. Nye explored in detail the rainbows occurring in the case of transmission of laser light through a water drop on a glass slide and reproduced the shapes of the generated rainbows using catastrophe theory.

More information about the meteorological rainbows can be found in the books of Van de Hulst [14] and Boyer [15], and in the review papers of Jackson [16] and Adam [17]. Interested readers can learn more about the optical rainbows generated with water drops from the book of Nye [18].

## 1.2 Particle Rainbows

### *1.2.1 Atomic Rainbows*

The atomic rainbow effect was discovered by Firsov [19]. He performed a thorough classical analysis of the process of elastic atomic scattering assuming that the interaction potential had both the repulsive and attractive components, and found that the deflection function, $\Theta(b)$, where $b$ is the impact parameter, had a minimum, $\Theta = \Theta_r$.[1] Consequently, the differential scattering cross section, $\sigma(\theta)$, where $\theta(b) = |\Theta(b)|$ is the scattering angle, tended to infinity at that point. This behavior is analogous to the behavior of the intensity of light scattered from a water droplet in the vicinity of the primary rainbow angle, as depicted in Fig. 1.3. A few years later, the effect was observed by Mason [20]. However, neither Firsov nor Mason connected it to the meteorological rainbow effect. The connection between the two effects was revealed by Ford and Wheeler [21, 22]. Using Airy's theory mentioned in Sect. 1.1, they gave a semiclassical theory of the atomic and nuclear scattering in the vicinity of an extremum of $\Theta(b)$, i.e., of the atomic and nuclear rainbow scattering. In the case of nuclear scattering, the effect they considered was the Coulomb rainbow effect, which will be considered in the next subsection.

The atomic rainbow effect was observed experimentally for the first time by Beck [23] with the crossed atomic and molecular beams at thermal kinetic energies.

[1] In Subsects. 1.2.1 and 1.2.2, the analysis is performed in the center-of-mass reference frame. This means that the considered variables are related to the relative particle, rather than to the projectile.

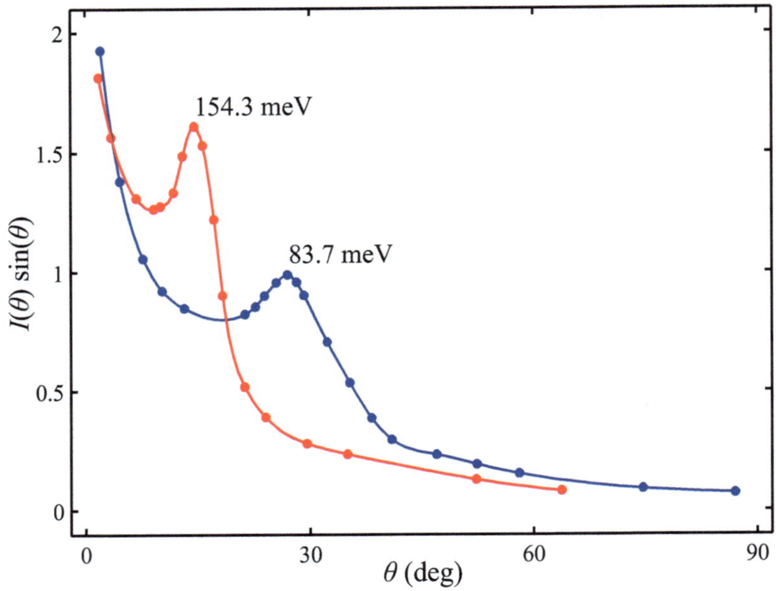


**Fig. 1.5** Experimental adjusted angular distributions of K atoms elastically scattered from HBr molecules at kinetic energies of 83.7 and 154.3 meV (Adapted from Ref. [23])

Figure 1.5 shows the dependences of the intensity of K atoms scattered from HBr molecules on $\theta$, $I(\theta)$, multiplied by $\sin\theta$ at kinetic energies of 83.7 and 154.3 meV obtained in that experiment. In each of the two cases, the maximum appearing in this adjusted angular distribution of scattered atoms corresponds to the primary rainbow. The region of $\theta$ below the maximum is the bright side of the rainbow, and the region of $\theta$ above the maximum is its dark side. Figure 1.6 gives the experimental adjusted angular distribution of $H^+$ ions scattered from Kr atoms at kinetic energy of 15.8 eV obtained by Weise et al. [24]. Here, one can see clearly five maxima of the distribution, which correspond to the primary rainbow and four associated supernumerary rainbows.

## 1.2.2 Coulomb and Nuclear Rainbows

As it has been said in the introductory part of this chapter, after the prediction of the rainbow effect in elastic atomic scattering, two rainbow effects were observed experimentally in elastic nuclear scattering. One of them, the Coulomb rainbow effect, appeared in the region of positive values of $\Theta(b)$, while the other one, the nuclear rainbow effect, occurred in the region of negative values of $\Theta(b)$[1]. The Coulomb rainbows were registered for the first time by Ford and Wheeler [21, 22] – in the previously obtained experimental results with α-particles of kinetic energy of

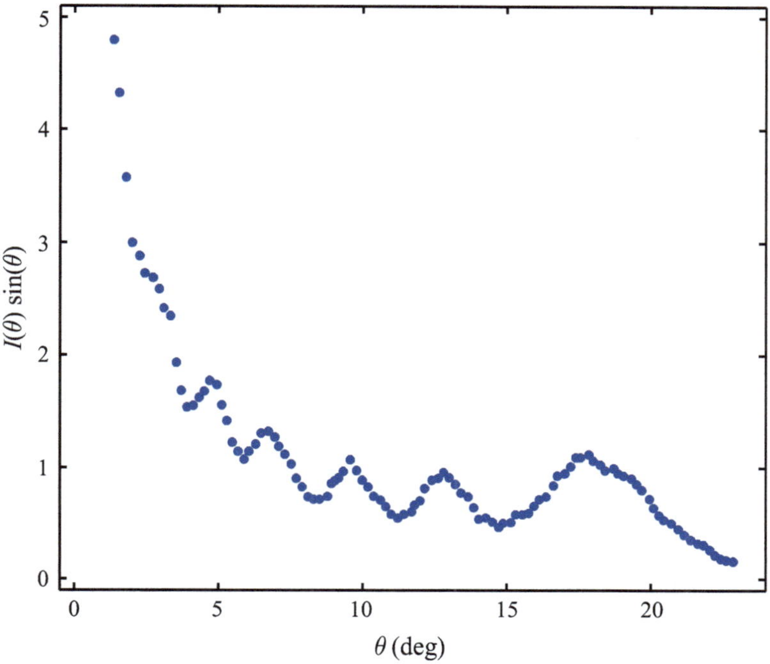


**Fig. 1.6** Experimental adjusted angular distribution of 15.8 eV $H^+$ ions elastically scattered from Kr atoms (Adapted from Ref. [24])

22 MeV and $^{107}$Ag nuclei [25], and 40 MeV α-particles and $^{181}$Ta and $^{232}$Th nuclei [26]. In fact, they registered only the quasiexponential decrease of $\sigma(\theta)$ for $\theta$ above the primary Coulomb rainbow angle. The nuclear rainbows were observed experimentally for the first time by Goldberg and Smith with 139 MeV α-particles and $^{58}$Ni nuclei [27].

Figure 1.7 shows the measured dependences $\sigma(\theta)$ divided by the corresponding differential cross sections for Rutherford scattering, $\sigma_R(\theta)$, in the cases of 1760 MeV $^{40}$Ar nuclei impinging on $^{60}$Ni, $^{120}$Sn, and $^{208}$Pb nuclei, i.e., the angular distributions of scattered nuclei, in the vicinity of the primary Coulomb rainbow angle [28]. The maxima and quasiexponential decrease of these dependences are due to the Coulomb rainbow effect.

Figure 1.8 gives the angular distribution of 104 MeV α-particles scattered from $^{40}$Ca nuclei [29, 30]. The oscillating character of the distribution is a consequence of the interference of its components corresponding to the positive and negative values of $\Theta(b)$[2]. The figure also contains these two components. The former

[2]In each of the angular distributions of scattered nuclei shown in Fig. 1.7, there are no such oscillations, because in the comprised region of $\theta$, its component corresponding to the positive values of $\Theta(b)$ is large compared to the component corresponding to the negative values of $\Theta(b)$.

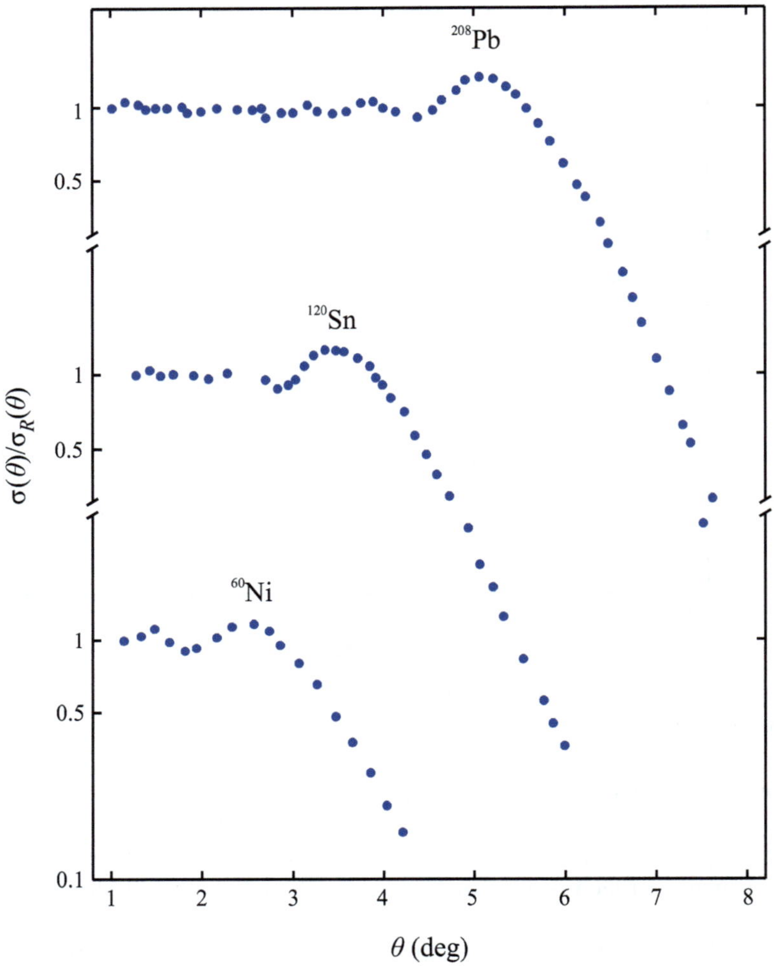


**Fig. 1.7** Experimental angular distributions of 1760 MeV $^{40}Ar$ nuclei elastically scattered from $^{60}Ni$, $^{120}Sn$, and $^{208}Pb$ nuclei in the vicinity of the primary Coulomb rainbow angle (Adapted from Ref. [28])

distribution decreases quasiexponentially, which is to be attributed to the Coulomb rainbow effect. The maxima and quasiexponential decrease clearly seen in the latter distribution are the signatures of the nuclear rainbow effect.

More information on the rainbows in elastic nuclear scattering can be found in the works of McVoy and Satchler [31] and Khoa et al. [32], and the references therein.

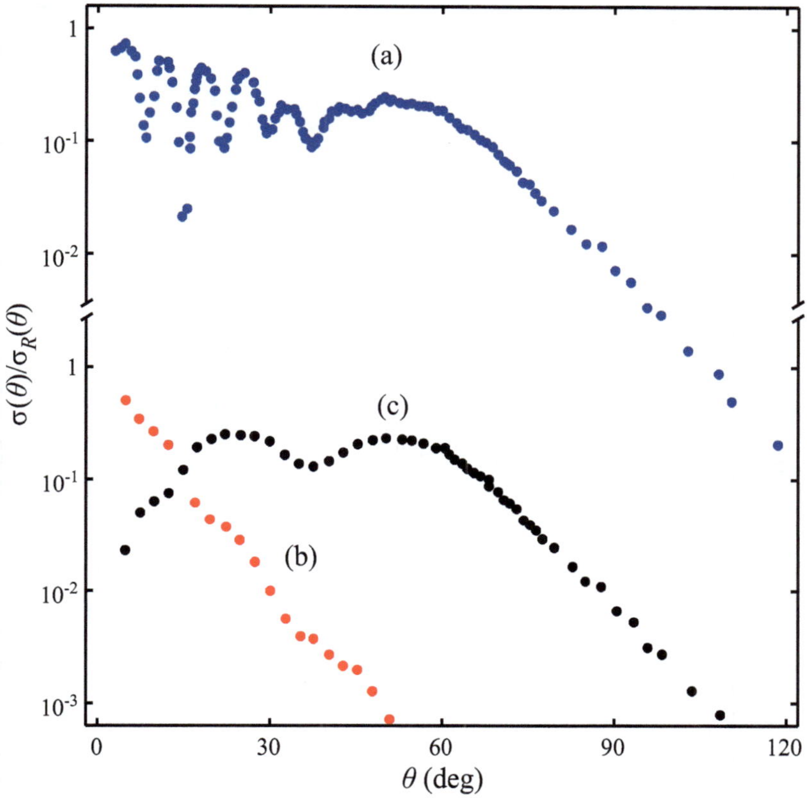


**Fig. 1.8** Experimental angular distributions of 104 MeV α-particles elastically scattered from $^{40}$Ca nuclei for (**a**) the positive and negative values of $\Theta(b)$, (**b**) the positive values of $\Theta(b)$, and (**c**) the negative values of $\Theta(b)$. $\sigma_R(\theta)$ is the corresponding differential cross section for Rutherford scattering (Adapted from Refs. [29, 30])

### *1.2.3 Surface Rainbows*

The surface rainbow effect was discovered by Oman [33] in a classical computer simulation of elastic scattering of Ne atoms of thermal kinetic energies from the (111) surface of a Ag crystal. Smith et al. [34] conducted the first measurements of the effect, with Ar atoms of thermal kinetic energies and the (001) surface of a LiF crystal. However, neither Oman nor Smith et al. connected the obtained results with the meteorological rainbow effect. That was done by McClure [35] in interpreting his classical computer experiments.

Figure 1.9 shows two experimentally obtained dependences of the intensity of Ne atoms of kinetic energy of 49 meV scattered from the (001) surface of a LiF crystal, $I$, on the scattering angle, $\theta_1$ [36]. The incoming and outgoing parts of each atom trajectory lied in a plane parallel to the (100) crystallographic plane, with the

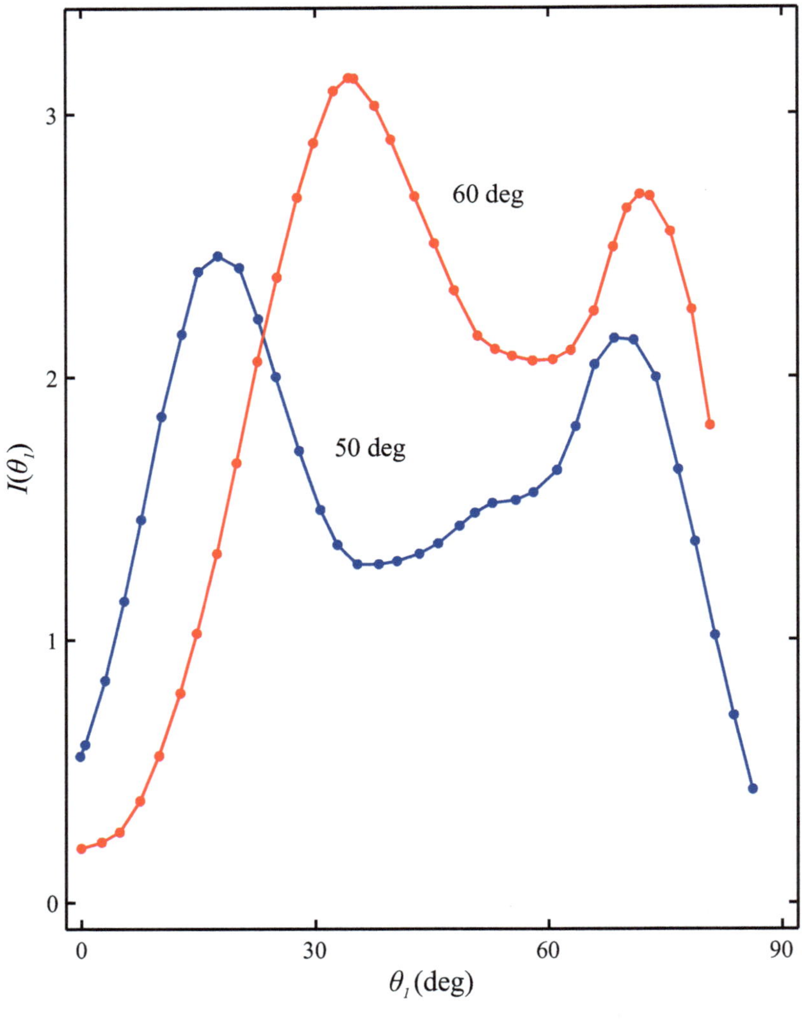


**Fig. 1.9** Experimental angular distributions of 49 meV Ne atoms elastically scattered from the (001) surface of a LiF crystal for $\theta_0$=50° and 60°. The incident atom velocity vectors are parallel to the (100) crystallographic planes (Adapted from Ref. [36])

incident angle, i.e., the angle between the incoming part of the trajectory and the surface normal, being $\theta_0$=50° or 60°, and with $\theta_1$ being the angle between the outgoing part of the trajectory and the surface normal. Each of these angular distributions of scattered atoms contains two pronounced maxima lying symmetrically relative to the angle of specular reflection, i.e., $\theta_1 = \theta_0$. These maxima correspond to two points between the adjacent atoms within an atomic string of the surface parallel to the <010> crystallographic direction. The authors found that, for the corresponding values of $\theta_1$, the classical differential cross section for elastic atom scattering tended to infinity. Hence, the maxima represent the primary

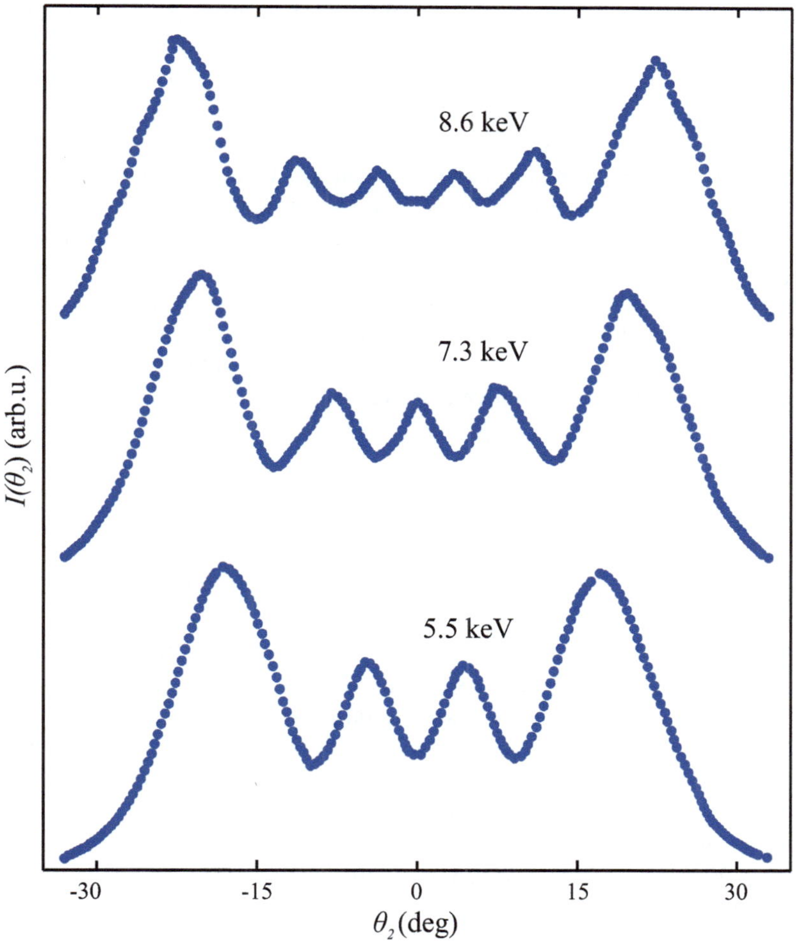


**Fig. 1.10** Measured angular distributions of 5.5, 7.3, and 8.6 keV He atoms scattered from the (001) surface of a LiF crystal for $\theta_0$=89.29°; the values of $I(\theta_2)$ are given in arbitrary units. The incident atom velocity vectors are parallel to the (010) crystallographic planes (Adapted from Ref. [39])

surface rainbow effect. The range of $\theta_1$ between the maxima is the bright side of the rainbow while the ranges of $\theta_1$ below the left maximum and above the right maximum are its dark sides.

Interested readers can learn more about the surface rainbows from the works of Kleyn and Horn [37] and Winter [38]. Figure 1.10 shows the experimental intensities of 5.5, 7.3, and 8.6 keV He atoms elastically scattered from the (001) surface of a LiF crystal, $I$, as functions of the scattering angle, $\theta_2$, obtained by Schüller and Winter [39] upon the publication of those two works. The incident atom velocity vectors were parallel to the (010) crystallographic planes, with the incident angle (relative to the surface normal) being $\theta_0$=89.29°. Thus, the atoms propagated along the <100> surface channels. The scattering angle of an atom ($\theta_2$) was the angle between its outgoing trajectory and the outgoing trajectory of the atom whose incoming trajectory lied in the plane containing the channel axis, and, as one can see in the figure, it could take positive and negative values. For each value of the kinetic energy, there are two more pronounced maxima lying in the peripheral parts of the angular distribution. The authors established that they corresponded to two lines between and parallel to the atomic strings defining the channel. These maxima occur due to the primary rainbow effect. In addition, there are two, three, and four less pronounced maxima lying between the more pronounced maxima for kinetic energies of 5.5, 7.3, and 8.6 keV, respectively. Their origin is the associated supernumerary rainbow effect. The appearance of these rainbows is attributed to the fact that the value of $\theta_0$ is very close to $\pi/2$, i.e., the value of the perpendicular component of the incident atom velocity vector is very small.

### *1.2.4 Crystal Rainbows*

The crystal rainbow effect was discovered by Nešković [40, 41] in a computer simulation of the transmission of protons of kinetic energy of 7 MeV through the <110> channel of a 150-nm thick Si crystal. Soon after that, the effect was observed experimentally by Krause et al. [42, 43], with 7 MeV protons and a 140-nm thick (100) Si crystal. The same group [44], from the Oak Ridge National Laboratory, Tennessee, USA, to be referred to as the Oak Ridge group, led by Datz and Moak, performed a series of additional measurements of the effect. The projectiles were 2–9 MeV protons and 6–30 MeV $C^{4+}$, $C^{5+}$, and $C^{6+}$ ions, and they were transmitted through the <100> channels of a 179-nm thick Si crystal. Between 2011 and 2014, a group from the National University of Singapore [45–48], to be referred to as the Singapore group, led by Breese, conducted a series of high-resolution measurements of the crystal rainbows with 2.0–0.7 MeV protons and a 55-nm thick (100) Si crystal.

Besides, Petrović et al. [49, 50] predicted the rainbow effects in 1 GeV proton channeling in a 1 μm long bundle of (10, 10) achiral single-wall carbon nanotube, and in 1 MeV positron channeling in a 200-nm long (11, 9) chiral single-wall carbon nanotube, respectively.

The rainbows occurring in transmission of charged particles through crystal channels and nanotubes will be considered in detail in Chaps. 3, 4, 5, and 6.

# Chapter 2
# Ion Channeling

## 2.1 Basic Facts

A positively charged particle impinging on a single crystal in a direction close to a major crystallographic axis will very probably experience a correlated series of many small angle collisions with the crystal's atoms [51]. As a result, an oscillatory particle motion in the space between the atomic strings of the crystal, i.e., along the axial crystal channels, is established. The process is called axial particle channeling. Under such conditions, the yields from close encounter processes involving the particle and crystal's atoms are very low. If the particle stays in one channel during the propagation through the crystal, the process is referred to as hyperchanneling. Similar behavior is expected if the incident particle direction is in a plane close to a major crystallographic plane, and the process is called planar channeling.

The possibility of channeling of positively charged particles in crystals was mentioned for the first time by Stark and Wendt in 1912 [52, 53]. However, the process was experimentally observed first about 50 years after that by Rol et al. [54], and then by Almén and Bruce [55]. Those authors showed that the yield of ion sputtering of monocrystalline target surfaces depended significantly on their orientation. In between the appearances of those two studies, Davies et al. [56, 57] reported on the anomalously long ranges of ions moving through polycrystalline targets. In a reaction to the results presented in [55], Robinson [58] emphasized the importance of the direction of ion penetration through crystals. Soon after that, Robinson and Oen [59, 60] reported on the computer simulations of slowing down of Cu atoms of kinetic energies between 1 and 10 keV in various monocrystalline targets, in which very large penetration depths of the ions with the incident directions close to the major crystallographic axes were observed. Those results were crucial for explaining the results of the above-mentioned three experimental studies – by the process of ion channeling. It became clear how an ion of a kinetic energy of the order of a keV or even MeV can move long in a potential well of a



N. Nešković et al., *Rainbows in Channeling of Charged Particles in Crystals and Nanotubes*, Lecture Notes in Nanoscale Science and Technology 25,
DOI 10.1007/978-3-319-61524-0_2

depth of the order of an eV. Lehman and Leibfried [61, 62], Nelson and Thompson [63], Lindhard [64], and Erginsoy [65] continued the theoretical exploration of the process. They introduced the concept of continuum interaction potential of a channeled ion and an atomic string and, using the approach of statistical mechanics, founded the analytical theory of ion channeling. The theory was described in detail by Lindhard [66].

The analytical theory of axial ion channeling is concentrated on a single atomic string and is based on the assumptions that: (i) the ion-string interaction is well described applying classical mechanics, (ii) the string can be treated as continuous, rather than discrete, and (iii) the transverse ion energy is conserved [66]. These assumptions enable an analytical description of the ion motion along the atomic string. In order to calculate the flux of channeled ions, one has to introduce an additional assumption – that the ion has an equal probability of finding itself at any point in the region of the transverse plane accessible to it, which is determined by its transverse energy [66]. However, this assumption, referred to as the assumption of statistical equilibrium in the transverse plane, makes the theory applicable only for thick crystals [67]. This theory has been improved by abandoning the assumption of conservation of the transverse ion energy, i.e., taking into account the energy losses of the ions in their collisions with the crystal's nuclei and electrons, and treating their motion along the atomic string as the diffusion in the transverse plane [68, 69].

## 2.2 Interaction Potentials

It is usually assumed that the interaction of an ion and an isolated atom can be sufficiently accurately described as a two-body interaction, i.e., the interaction of the nuclei of the collision partners screened by their electron clouds [51]. One also supposes that the ion-atom interaction potential does not depend on their relative velocity and that it is not influenced by the inelastic processes during the encounter. The interaction of an ion and a crystal is most often treated starting from such an interaction potential. This approach is applied in studying the process of charged particle channeling as well.

It has been demonstrated that the (statistical) Thomas-Fermi atomic model can lead to sufficiently accurate atom-atom interaction potentials [51]. With this approximation, the interaction potential reads

$$U(r_a) = \frac{Z_1 Z_2 e^2}{r_a} \varphi_s\left(\frac{r_a}{a_s}\right), \tag{2.1}$$

where $Z_1$ and $Z_2$ are the projectile and target atomic numbers, respectively, $e$ is the elementary charge, $r_a$ is the atom-atom distance, $\varphi_s(r_a/a_s)$ is the atom-atom screening function, and $a_s$ is the atom-atom screening length. The function $\varphi_s$ describes the screening of the nuclei of the collision partners with their electron clouds, but without taking care of the shell structures of the clouds. It is given as the numerical solution of a differential equation [70]. Molière [71] found a good analytical approximation of the screening function,

$$\varphi_{\mathrm{M}}\left(\frac{r_a}{a_s}\right)=\sum_{\ell=1}^{3}\alpha_\ell \exp\left(-\frac{\beta_\ell r_a}{a_s}\right), \tag{2.2}$$

where $(\alpha_\ell)$=(0.10, 0.55, 0.35) and $(\beta_\ell)$=(6.0, 1.2, 0.3) are the fitting parameters. This expression has been commonly used in treating atomic collisions in solids.

Lindhard [66] proposed another approximation of the Thomas-Fermi atom-atom screening function, which reads

$$\varphi_{\mathrm{L}}\left(\frac{r_a}{a_s}\right)=1-\frac{1}{\left[1+C_{\mathrm{L}}(r_a/a_s)^{-2}\right]^{1/2}}, \tag{2.3}$$

where $C_{\mathrm{L}}$ is a fitting parameter whose standard value is 3. In this expression, the two terms represent the nuclear and electronic contributions to the atom-atom interaction potential, respectively. This screening function has been widely used in studying the process of ion channeling.

In the Thomas-Fermi atomic model, the atomic screening length is

$$a_{\mathrm{TF}}=\left(\frac{9\pi^2}{128}\right)^{1/3} Z_2^{-1/3} a_0, \tag{2.4}$$

where $a_0=\hbar^2/(m_e e^2)$=0.0529 nm is the Bohr radius, $m_e$ is the electron mass, and $\hbar$ is the reduced Planck constant [72]. Firsov [73] showed that, when an atom-atom collision is considered, one should use the screening length

$$a_{\mathrm{F}}=\left(\frac{9\pi^2}{128}\right)^{1/3}\left(Z_1^{1/2}+Z_2^{1/2}\right)^{-2/3} a_0. \tag{2.5}$$

This expression reduces to Eq. (2.4) when $Z_1 \ll Z_2$. An alternative expression for the atom-atom screening length,

$$a_{\mathrm{L}}=\left(\frac{9\pi^2}{128}\right)^{1/3}\left(Z_1^{2/3}+Z_2^{2/3}\right)^{-1/2} a_0, \tag{2.6}$$

was proposed by Lindhard [72]. This means that for $a_s$, appearing in Eqs. (2.2) and (2.3), one should choose $a_{\mathrm{TF}}$, $a_{\mathrm{F}}$, or $a_{\mathrm{L}}$. It has been established [51] that, when the projectile is a bare nucleus, the most appropriate choice is $a_{\mathrm{TF}}$. If the projectile is a screened nucleus, one should choose $a_{\mathrm{F}}$ or $a_{\mathrm{L}}$.

Ziegler, Biersack, and Littmark (ZBL) [74–76] introduced an approximation of the screening function based on the Hartree-Fock method and the assumption that the shell structures of the electron clouds of the partners did not change during the collision process. They applied it to 261 atomic pairs. This approximation, which is designated as the universal atom-atom screening function, is often used in treating atomic collisions. It reads

$$\varphi_{\mathrm{ZBL}}\left(\frac{r_a}{a_{\mathrm{ZBL}}}\right) = \sum_{\ell=1}^{4} \gamma_\ell \exp\left(-\frac{\delta_\ell r_a}{a_{\mathrm{ZBL}}}\right), \tag{2.7}$$

with

$$a_{\mathrm{ZBL}} = \left(\frac{9\pi^2}{128}\right)^{1/3} \left(Z_1^p + Z_2^p\right)^{-1} a_0 \tag{2.8}$$

as the atom-atom screening length, where $(\gamma_\ell)$=(0.1818, 0.5099, 0.2802, 0.02817), $(\delta_\ell)$=(3.2, 0.9423, 0.4028, 0.2016), and $p$=0.23 are the fitting parameters.

## 2.3 Computer Simulations

### 2.3.1 *Applicability of Classical Mechanics*

Let us imagine a collision of two bare nuclei, the projectile and target, with atomic numbers $Z_1$ and $Z_2$, whose masses are $m_1$ and $m_2$, respectively, with the condition $m_2 >> m_1$. The projectile velocity and impact parameter are $v$ and $b$, respectively. Initially, the target is at rest and, being much heavier than the projectile, can be taken to stay that way. The relativistic effects are ignored. We treat the projectile as a wave packet. According to Bohr [77], there are two contributions to the uncertainty of the projectile scattering angle. The first contribution is due to the uncertainty of its initial position while the source of the second contribution is its diffraction. The resulting two conditions reduce to Bohr's condition for application of classical mechanics, which reads

$$\frac{2Z_1Z_2e^2}{\hbar v} >> 1, \tag{2.9}$$

where $e$ is the elementary charge and $\hbar$ is the reduced Planck constant. According to this expression, there is a projectile kinetic energy above which classical mechanics cannot be used.

Lindhard [66] showed that, when two screened nuclei collide and the corresponding screening function is given by Eq. (2.3), the condition for application of classical mechanics is

$$\frac{2Z_1Z_2e^2}{\hbar v} >> 1 + \frac{3b^2}{C_{\mathrm{L}} a_s^2}. \tag{2.10}$$

This means that for a sufficiently high ion kinetic energy or a sufficiently large ion impact parameter, i.e., a sufficiently small ion scattering angle, quantum mechanics must be used. For example, if a proton of kinetic energy of 2 MeV is incident on a Si atom, $C_L$=3, and $b = a_s$, the left-hand side of the inequality (2.10) is ~3 and its right-hand side is 2, i.e., the inequality is not valid. Thus, since in ion channeling $b$ is above ~$a_s$, making, in this case, the right-hand side of the inequality even larger than 2, one might conclude that classical mechanics cannot be used for describing the process of 2 MeV proton channeling in an axial channel of a Si crystal. However, Lindhard [66] demonstrated that, even though an individual small angle collision may not be amenable to classical treatment, the situation is opposite for a correlated series of such collisions. He did that in cases of a real (discrete) atomic string of a crystal, consisting of a number of equally displaced atoms, and of a continuous string, obtained after a uniform longitudinal smearing of its atoms. In case of the real atomic string, the uncertainty of the ion scattering angle does not grow during the series of collisions. Also, the necessity to use quantum mechanics decreases as the ion kinetic energy increases. In case of the continuous atomic string, Lindhard obtained the following condition for application of classical mechanics:

$$2\left(\frac{Z_1 Z_2 m_1 a_s^2}{m_e d a_0}\right)^{1/2} >> 1, \tag{2.11}$$

where $m_e$ is the electron mass, $d$ is the distance between the atoms of the atomic string, and $a_0$ is the Bohr radius. This condition is independent of the projectile velocity and is certainly satisfied for $m_1 >> m_e$.

As it has been said in the Prologue, we shall discuss here about ion channeling in various crystals and carbon nanotubes, and about positron channeling in carbon nanotubes. In accordance with the inequality (2.11), the cases of ion channeling will be described using classical mechanics, and the cases of positron channeling using quantum mechanics [66].

### 2.3.2 Continuum String Model

From a classical point of view, a positively charged particle is considered to be channeled with respect to an atomic string of a crystal if it is incident on the string at an angle, $\varphi$, small enough that it is scattered away from the string by a correlated series of many glancing collisions with the atoms of the string [51]. As $\varphi$ becomes greater than

$$\psi_c = \left(\frac{2Z_1 Z_2 e^2}{dE}\right)^{1/2}, \tag{2.12}$$

where $E$ is the particle kinetic energy, referred to as the critical angle for axial channeling, for which the distance of closest approach of the particle to the atomic string is ~$a_s$ [66], the particle begins to approach the string so closely that it no longer remains channeled. Instead, it begins to interact dominantly with the individual atoms and is scattered away from the atomic string upon one or a few larger angle collisions, i.e., it becomes dechanneled. For $\varphi >> \psi_c$, one can take that the particle moves randomly with respect to the atomic string. Thus, in the process of axial particle channeling, it is central that the particle interaction with each atomic string includes many consecutive small angle collisions and no collisions with the individual atoms. Under such conditions, one can use the continuum string model, i.e., assume that the particle interacts with a continuous atomic string, which is obtained after a uniform longitudinal smearing of its atoms [61–66]. In this case, it is also said that the continuum approximation is employed.

Let us take that the $z$ axis of the reference frame is the longitudinal axis, and the $x$ and $y$ axes are the transverse axes, being the vertical and horizontal axes, respectively. The atomic strings are parallel to the $z$ axis, with the median plane of the crystal coinciding with the $xy$ plane. The thermal vibrations of the atoms of the strings are ignored. The continuum interaction potential of the particle and the $i$th atomic string is defined as

$$V_i(\rho_i) = \frac{1}{d}\int_{-\infty}^{+\infty} U\left[\left(\rho_i^2 + z^2\right)^{1/2}\right]dz, \tag{2.13}$$

where $U$ is given by Eq. (2.1), $\rho_i^2 = (x - x_i)^2 + (y - y_i)^2$ is the particle-string distance squared, $x$ and $y$ are the transverse components of the particle position vector, $x_i$ and $y_i$ are the transverse coordinates of the string, and $z$ is the longitudinal component of the particle position vector [51]. Hence, one obtains

$$V_i(\rho_i) = \frac{2Z_1Z_2e^2}{d} f_s\left(\frac{\rho_i}{a_s}\right), \tag{2.14}$$

with

$$f_s(\xi_i) = \int_0^{\infty} \left(\xi_i^2 + \zeta^2\right)^{-1/2} \varphi_s\left[\left(\xi_i^2 + \zeta^2\right)^{1/2}\right] d\zeta \tag{2.15}$$

and $\xi_i = \rho_i/a_s$. For Molière's atom-atom screening function, given by Eq. (2.2),

$$f_s(\xi_i) = \sum_{\ell=1}^{3} \alpha_\ell K_0(\beta_\ell \xi_i), \tag{2.16}$$

where $K_0$ denotes the zero-order modified Bessel function of the second kind [51]. For Lindhard's atom-atom screening function, defined by Eq. (2.3),

$$f_s(\xi_i) = \ln\left(1 + \frac{C_L}{\xi_i^2}\right)^{1/2} \tag{2.17}$$

[66]. Finally, for the ZBL atom-atom screening function, determined by Eqs. (2.7) and (2.8),

$$f_s(\xi_i) = \sum_{\ell=1}^{4} \gamma_\ell K_0(\delta_\ell \xi_i) \tag{2.18}$$

[51]. The particle-string continuum interaction potential can be regarded as the zero-order term of the Fourier expansion of the real (periodic) interaction potential [51]. The coefficients of the remaining terms of the real interaction potential decrease rapidly with the particle distance from the atomic string [61, 62]. The particle-crystal continuum interaction potential is the sum of the particle-string interaction potentials.

Usually, the particle velocity is such that the time of its interaction with an atom of a string is much shorter than the period of thermal vibrations of the atom [51]. This means that, in the case in question, the atomic string can be treated as stationary. Thus, the particle in fact experiences the continuum potential of the atomic string averaged over the thermal displacements of its atoms from their equilibrium positions. The resulting particle-string continuum interaction potential is obtained as it was shown by Appleton et al. [78]. In such a derivation, it is assumed that the atoms of the string vibrate independently, and that, for each translational degree of freedom, the probability of an atomic displacement is described by a Gaussian distribution function. The standard deviation of this distribution, i.e., the one-dimensional atomic thermal vibration amplitude, $\sigma_{\text{th}}$, is determined using the Debye theory of thermal vibrations [79]. The obtained thermally averaged interaction potential begins to deviate considerably from the corresponding static interaction potential when the particle distance from the atomic string becomes comparable to $\sigma_{\text{th}}$ [51].

If the particle energy losses in its collisions with the crystal's nuclei and electrons, i.e., its nuclear and electronic energy losses, can be neglected, the longitudinal component of the particle velocity vector can be treated as constant. Thus, if the continuum string model is used, only the transverse particle motion needs to be treated.

Let us assume that the process of axial ion channeling we are interested in can be accurately described using classical mechanics and the continuum string model. A computer simulation of the process requires that we select: (i) an ion-atom interaction potential, among those given in Sect. 2.2, (ii) a one-dimensional atomic thermal vibration amplitude, (iii) an adequate expression for the ion nuclear energy loss rate, (iv) adequate expressions for the ion electronic energy loss rate and the rate of the corresponding uncertainty of the ion scattering angle, and (v) an uncertainty of the ion incident angle, corresponding to the incident ion beam divergence [80].

Then, the transverse components of the initial ion position vector, in the entrance plane of the crystal, are chosen from a (two-dimensional) uniform distribution over the region of a crystal channel. The choosing is done randomly, by a random number generator, or uniformly. In doing this, we avoid the initial position vectors lying within the small circular regions around the atomic strings defining the channel since the resulting channeling process would obviously include larger angle collisions. These regions are limited by the ion-atom screening length, $a_s$. Upon that, the transverse components of the initial ion velocity vectors are chosen from a two-dimensional Gaussian distribution defined by the selected uncertainty of the ion incident angle. This choosing is also performed randomly or uniformly.

With the initial conditions for the ion propagation along the channel defined, its equations of motion in the transverse plane are solved, and its trajectory is obtained. At the end of each step of the calculation, the longitudinal component of the ion velocity vector is changed on the basis of the calculated ion nuclear and electronic energy losses during the step, whereas the transverse components of the ion velocity vector are randomly changed on the basis of the calculated uncertainties of the components of the ion scattering angle during the step. In most cases, the nuclear energy loss can be disregarded [66].

When the ion equations of motion in the transverse plane are solved, one can calculate the transverse components of the final ion position vector, in the exit plane of the crystal, and the components of the final ion channeling angle, i.e., the ion transmission angle. The spatial and angular distributions of transmitted ions are generated with a sufficiently large number of incident ions. If the transverse components of the initial ion position and velocity vectors are chosen randomly, one says that the distributions are obtained using the Monte Carlo method.

The spatial and angular distributions of transmitted ions generated following such a computer simulation approach will be shown and analyzed in Chaps. 3, 4 and 5. In all these cases, the nuclear energy loss will be neglected [66].

### 2.3.3 Binary Collision Model

In the binary collision model of ion channeling in a crystal, the crystal is treated as real, i.e., a three-dimensional symmetric arrangement of atoms. It is assumed that the ion has only one important interaction at a time as it moves through the crystal. In each collision, the ion scattering angle is computed using the momentum approximation [61, 62], or this is done by numerical integration. Two computer simulation procedures for using this model have been developed. In both of them, it is assumed that the condition for application of classical mechanics is satisfied. In the former procedure [59, 60], the unit cell of the crystal and its translational properties are defined, and the computer code is left to search and find the crystal's atom with which the next ion interaction will happen. In the latter procedure

[81, 82], the crystal is viewed as composed of the symmetrically arranged atomic strings, consisting of the equally displaced atoms. In this case, the next important ion collision occurs with the next atom of the atomic string currently involved in the channeling process, referred to as the current central string. The critical moment in the calculation is the one at which the attention is transferred from the current central atomic string to the next one. The latter procedure has proven to be more efficient than the former one [82]. Often, it is necessary to include the contributions of the surrounding atomic strings to the ion scattering angle. This is done by treating these atomic strings as continuous, i.e., employing the continuum approximation.

In a computer simulation going on according to the above-described latter procedure, the ion trajectory is a straight line between the equidistant transverse planes with the spacing equal to the distance between the atoms of the current central atomic string. In each of these planes, the components of the ion position and velocity vectors are updated on the basis of the outcomes of the four events [83]: (i) the ion collision with the nearest atom of the string, being a screened nucleus, taking into account its thermal displacement from the equilibrium position, (ii) the ion scattering from the surrounding atomic strings governed by their thermally averaged continuum interaction potentials, (iii) the ion energy loss due to its collision with the nearest screened nucleus, and (iv) the ion energy loss and the changes of its scattering angle due to its collisions with the crystal's electrons. As it has been said in the previous subsection, in most cases, the nuclear energy loss can be neglected [66]. If the updated ion velocity vector makes a large angle with the atomic string, it is assumed that the ion motion has become random. For the ion-atom interaction potential, one of those mentioned in Sect. 2.2 is selected.

The initial conditions for the ion motion along the channel are chosen in the same way as in the computer simulation employing the continuum string model, which has been described in the previous subsection. One may also include in the calculations the mosaic spread of the crystal, the disorder of the crystal surface, the correlations of the thermal vibrations of the adjacent atoms of the current central string, and the changes of the ion charge during the propagation.

The output quantities in such a computer simulation code may include: (i) the probability that an ion experiences a sufficiently close encounter with a crystal's atom to induce an event such as a large angle scattering, an energetic atom recoiling, or a nuclear reaction, (ii) the relative number of recoiled atoms having the kinetic energy above a certain value, and (iii) the spatial and angular distributions of transmitted ions. Two such codes that have been used most so far are the LAROSE code, made by Barrett [81, 82], and the FLUX code, developed by Smulders and Boerma [83]. In Subsect. 3.1.2, we shall present and analyze the angular distributions generated with the former code.

# Chapter 3
# Crystal Rainbows

This chapter is devoted to the process of ion transmission through axial crystal channels [84, 40–43, 85–88, 44, 89–100, 80, 101, 102]. The scheme of the process is shown in Fig. 3.1. In accordance with what has been said in Sect. 2.1, the motion of an ion close to the channel axis can be treated as an oscillatory motion around the axis. If the impact parameter vector of the ion and the angle between its incident momentum vector and the channel axis are fixed, the number of oscillations the ion makes before leaving the crystal depends on its incident kinetic energy [98]. When this energy is sufficiently high for the ion to make less than about a quarter of an oscillation, its trajectory can be approximated by a straight line. If this is true for the majority of ions, one says that the crystal is very thin. When, however, the majority of ions make between about a quarter of an oscillation and about one oscillation, the crystal is thin. The crystal is said to be thick if the majority of ions make more than about one oscillation. Finally, if the majority of ions make much more than one oscillation, the crystal is very thick.

## 3.1 Ion Channeling in Very Thin Crystals

### *3.1.1 Model of Crystal Rainbows*

Let us now consider an ion propagating along an axial channel of a very thin crystal [40, 41]. As it is shown in Fig. 3.1, the $z$ axis of the reference frame, being the longitudinal axis, coincides with the channel axis, and the origin lies in the median plane of the crystal. The $x$ and $y$ axes of the reference frame, being the transverse axes, are the vertical and horizontal axes, respectively. The incident ion momentum vector is parallel to the channel axis. In this case, one can apply the momentum approximation [61]. Within this approximation, the $x$ and $y$ components of the ion transmission angle are



N. Nešković et al., *Rainbows in Channeling of Charged Particles in Crystals and Nanotubes*, Lecture Notes in Nanoscale Science and Technology 25,
DOI 10.1007/978-3-319-61524-0_3

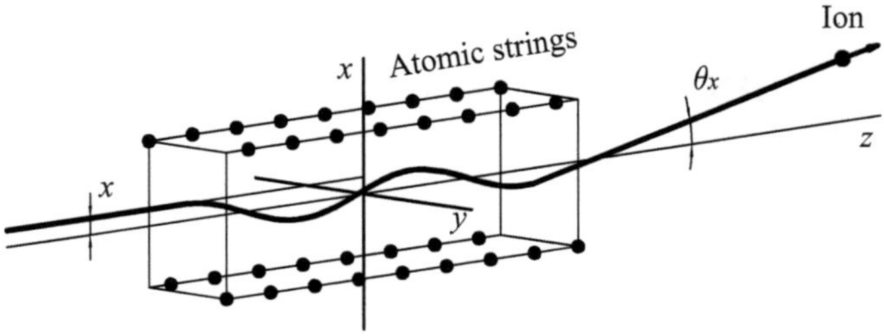


**Fig. 3.1** Scheme of a process of ion transmission through an axial crystal channel [101]. The $z$ axis coincides with the channel axis, and the origin lies in the median plane of the crystal. The channel is defined by four atomic strings, and the incident ion momentum vector is parallel to the $z$ axis. The $x$ components of the ion impact parameter vector and the transmission angle are denoted by $x$ and $\theta_x$, respectively. In this case, the $y$ components of these variables are equal to zero

$$\theta_x = -\frac{1}{2E}\sum_{i=1}^{M}\sum_{k=1}^{N}\int_{-\infty}^{+\infty}\partial_x U_{ik} dz \tag{3.1}$$

and

$$\theta_y = -\frac{1}{2E}\sum_{i=1}^{M}\sum_{k=1}^{N}\int_{-\infty}^{+\infty}\partial_y U_{ik} dz, \tag{3.2}$$

respectively, where $E$ is the incident ion kinetic energy, $U_{ik} = U_{ik}(x, y, z)$ is the interaction potential of the ion and $k$th atom of the $i$th atomic string of the crystal, $x$ and $y$ are the transverse components of the initial ion position vector, i.e., the components of its impact parameter vector, $z$ is the longitudinal component of the ion position vector along the channel, $M$ is the number of atomic strings of the crystal, $N$ is the number of atoms in one string, $\partial_x \equiv \partial/\partial_x$, and $\partial_y \equiv \partial/\partial_y$. These expressions show that, from the point of view of $\theta_x$ and $\theta_y$, the $(MN + 1)$ particle collision in question reduces to $MN$ independent two-particle collisions, i.e., the effect of multiple scattering cannot be observed. It is clear that the contributions of the atoms of one atomic string to $\theta_x$ and $\theta_y$ do not depend on their positions within the string. Consequently, in accordance with Eq. (2.13),

$$\theta_x = -\frac{L}{2E}\sum_{i=1}^{M}\partial_x V_i \tag{3.3}$$

and

$$\theta_y = -\frac{L}{2E}\sum_{i=1}^{M}\partial_y V_i, \tag{3.4}$$

where $L = Nd$ is the thickness of the crystal, $d$ is the distance between the atoms of a string, and $V_i = V_i(x, y)$ is the continuum interaction potential of the ion and $i$th string [66]. These expressions demonstrate that, from the point of view of $\theta_x$ and $\theta_y$, the process we analyze in fact consists of $M$ independent two-particle collisions. Thus, by applying the momentum approximation, we neglect the fact that the atoms of the atomic strings are arranged in a nonrandom way, i.e., the effect of longitudinal correlations. In other words, we implicitly introduce the continuum approximation, in which the atomic strings are treated as if they are continuous, rather than discrete [66].

In the case we analyze, the ion differential transmission cross-section reads

$$\sigma = \frac{1}{|J|}, \tag{3.5}$$

where

$$J = \partial_x\theta_x\partial_y\theta_y - \partial_y\theta_x\partial_x\theta_y \tag{3.6}$$

is the Jacobian of the functions $\theta_x(x, y)$ and $\theta_y(x, y)$ [103, 104], which is the ratio of the infinitesimal surfaces in the transmission angle (TA) plane and the impact parameter (IP) plane. The variable $J$ can be expressed as

$$J = J_0 + J', \tag{3.7}$$

where

$$J_0 = \sum_{i=1}^{M} J_{ii}, \quad J' = \sum_{i=1}^{M-1}\sum_{j=i+1}^{M}\left(J_{ij} + J_{ji}\right) \tag{3.8}$$

and

$$J_{ij} = \left(\frac{L}{2E}\right)^2\left(\partial_{xx}V_i\partial_{yy}V_j - \partial_{xy}V_i\partial_{xy}V_j\right), \tag{3.9}$$

where $\partial_{xx} \equiv \partial^2/\partial x^2$, $\partial_{yy} \equiv \partial^2/\partial y^2$, and $\partial_{xy} \equiv \partial^2/(\partial x\partial y)$. The term $J_{ii}$ describes the ion scattering from the $i$th atomic string, while the sum of terms $J_{ij}$ and $J_{ji}$ describes the coupling between the $i$th and $j$th string as seen by the ion. Hence, from the point of view of $\sigma$, the $(MN + 1)$ particle collision under consideration does not reduce to $MN$ independent two-particle collisions, i.e., the effect of multiple scattering can be observed. In this case, the process we study is composed of $M$ nonindependent two-particle collisions. This is a classical interference effect – the contributions of the atomic strings to $\sigma$ interfere.

One can see that the equation $J'=0$ defines the lines along which the effect of multiple scattering is not observable. These lines are called the single-scattering lines. The rainbow lines, i.e., the lines along which $\sigma$ is infinite, are determined by the equation $J=0$. All these lines are defined in the IP plane and the TA plane. The effect determined by them is called the crystal rainbow effect, while this model of ion transmission through axial channels of very thin crystals is referred to as the model of crystal rainbows. The explanation of the effect via the zeros of the function $J(x, y)$ is its mathematical explanation.

It should be mentioned that the above-described approach to classical small-angle scattering from many-particle targets was applied for the first time in solving the problem of scattering of a charged particle from an electric dipole [84]. The problem appeared in the analysis of the experimental results related to ion scattering from metal surfaces covered with alkali atoms [105, 106].

Let us now apply the model of crystal rainbows to the transmission of protons through the <110> channel of a very thin Si crystal. The values of $E$ and $L$ will be 7 MeV and 150 nm, respectively. This means that the number of atoms in one atomic string is $N=392$. We assume that the ion-atom interaction potential is obtained using the Thomas-Fermi atomic model and that it can be well described by Lindhard's expression with the Thomas-Fermi atomic screening length, by Eqs. (2.1), (2.3), and (2.4),

$$U_{\rm L}=\frac{Z_1Z_2e^2}{r_a}\left[1-\frac{r_a}{\left(r_a^2+C_{\rm L}a_{\rm TF}^2\right)^{1/2}}\right], \tag{3.10}$$

where $Z_1=1$ and $Z_2=14$ are the atomic numbers of the ion and atom, respectively, $e$ is the elementary charge, $r_{\rm a}$ is the ion-atom distance, $a_{\rm TF}$ is the screening length, and $C_{\rm L}=3.00$ [66]. As a result, according to Eqs. (2.13), (2.14), (2.15), and (2.17), the continuum interaction potential of the ion and $i$th atomic string reads

$$V_{{\rm L}i}=\frac{Z_1Z_2e^2}{d}\ln\left(1+\frac{C_{\rm L}a_{\rm TF}^2}{\rho_i^2}\right), \tag{3.11}$$

where $d=0.543082$ nm [107], $\rho_i^2=(x-x_i)^2+(y-y_i)^2$ is the ion-string distance squared, and $x_i$ and $y_i$ are the transverse coordinates of the string. The variable $V_{{\rm L}i}/(Z_1e)$ is the continuum potential of the $i$th atomic string. The number of atomic strings is $M=20$, i.e., we take into account the strings lying on the four coordination circles nearest to the channel axis. We do not take into account the thermal vibrations of the crystal's atoms and the ion collisions with the crystal's electrons.

In this case, there are one single-scattering line and one rainbow line in the IP plane. These two lines are shown in Fig. 3.2. One can see that the single-scattering line connects the regions of the IP plane in which the transmission process is dominated by the interaction with one atomic string. This line is "attracted" by the atomic strings defining the channel. On the other hand, the rainbow line lies in the region of the IP plane where the contributions of more atomic strings are important. This line is "repelled" by the atomic strings defining the channel. It must be emphasized that the coordinates of the points of these two lines are

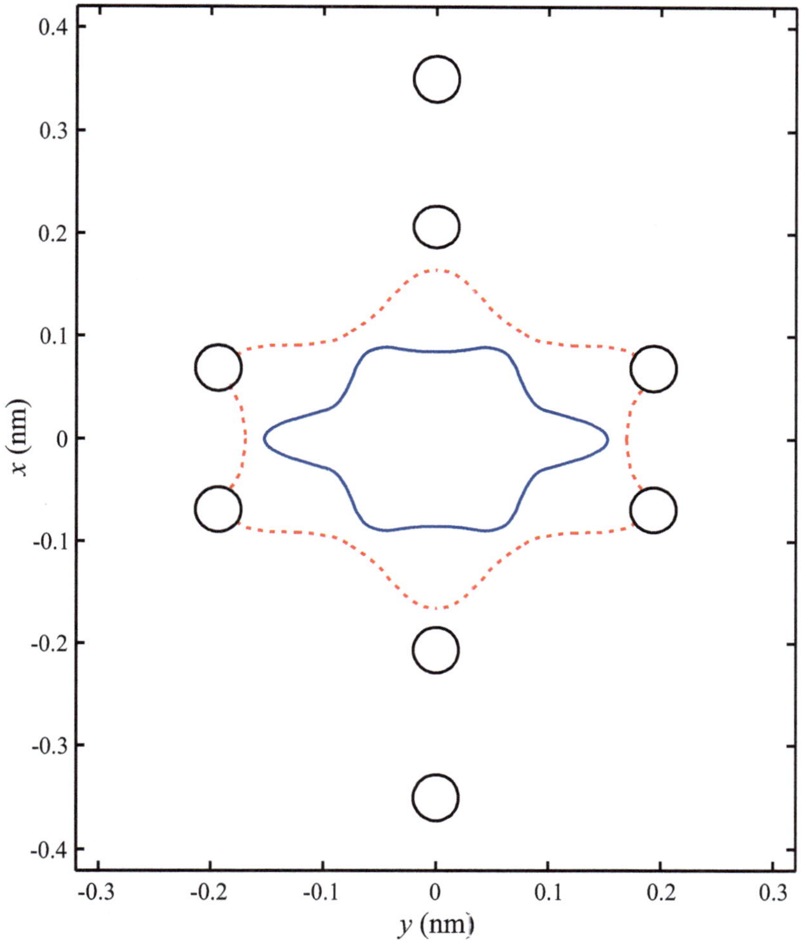


**Fig. 3.2** Rainbow line (*solid line*) and single-scattering line (*dashed line*) in the IP plane for protons transmitted through the <110> channel of a Si crystal [41]. *Eight small open circles* represent the atomic strings lying on the first and second coordination circles

determined solely by the arrangement of atomic strings and their continuum potential – they do not depend on $Z_1$, $E$, and $L$. Figures 3.3 and 3.4 give the dependences of $\sigma$ on the variable $\rho = (x^2 + y^2)^{1/2}$ for $\phi = \tan^{-1}(y/x) = 30$ and 210°, and on $\phi$ for $\rho = 0.085$ nm, respectively. The figures also contain the corresponding dependences of $\sigma$ when the term $J'$ is not taken into account, i.e., when the effect of multiple scattering is disregarded. The radial dependence of $\sigma$ with the effect of multiple scattering taken into account tends to infinity at two points, which belong to the rainbow line in the IP plane. However, the radial dependence of $\sigma$ with the effect of multiple scattering neglected is smooth. Thus, the two radial dependences clearly show that the crystal rainbow effect occurs due to the effect of multiple scattering, i.e., because the contributions of the atomic strings to $\sigma$ interfere. This is

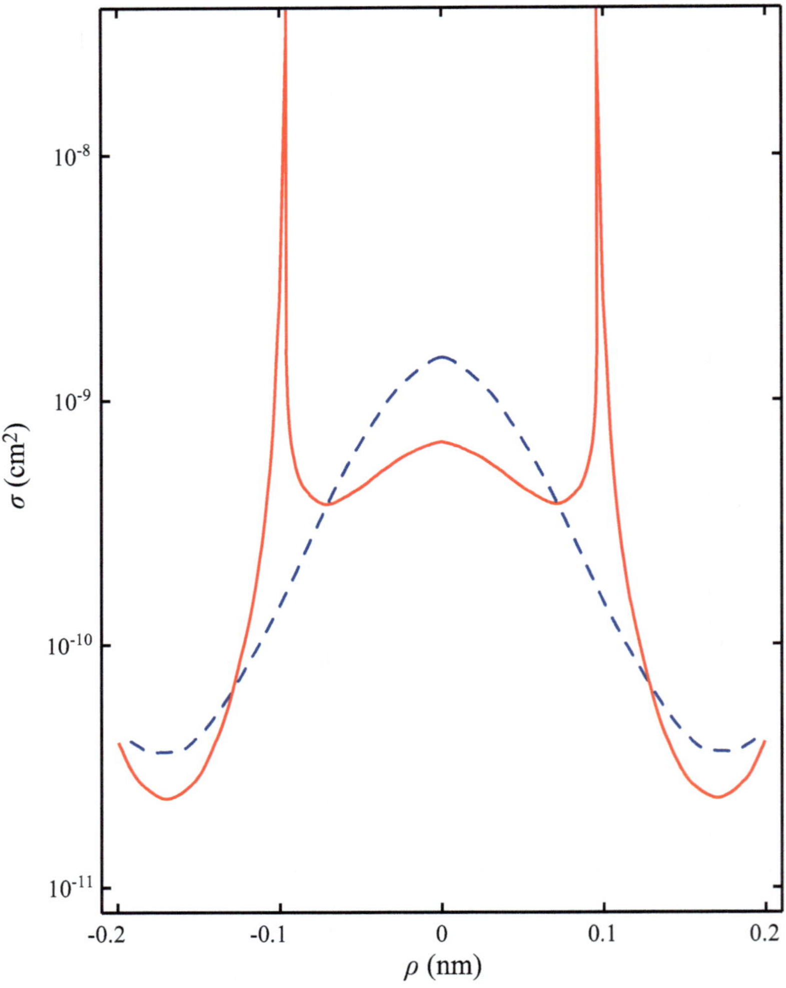


**Fig. 3.3** Dependences of the proton differential transmission cross section in the case of 7 MeV protons and the <110> channel of a 150-nm thick Si crystal on $\rho$ for $\phi = 30^{\circ}$ and 210° without (*dashed line*) and with (*solid line*) the effect of multiple scattering [41]

a manifestation of the fact that the crystal is more than a simple sum of the atomic strings. Hence, the effect is a gestalt or synergistic effect. These conclusions constitute the physical explanation of the crystal rainbow effect. The azimuthal dependences of $\sigma$ demonstrate that, when the effect of multiple scattering is neglected, this variable is maximal between the atomic strings and minimal toward them, and, when the effect is included, it is the opposite. The rainbow line in the TA plane is depicted in Fig. 3.5. It contains six cusps pointing to the atomic strings lying on the first coordination circle. This line divides the TA plane into the bright and dark regions – the bright region is on the inner side of the rainbow while the

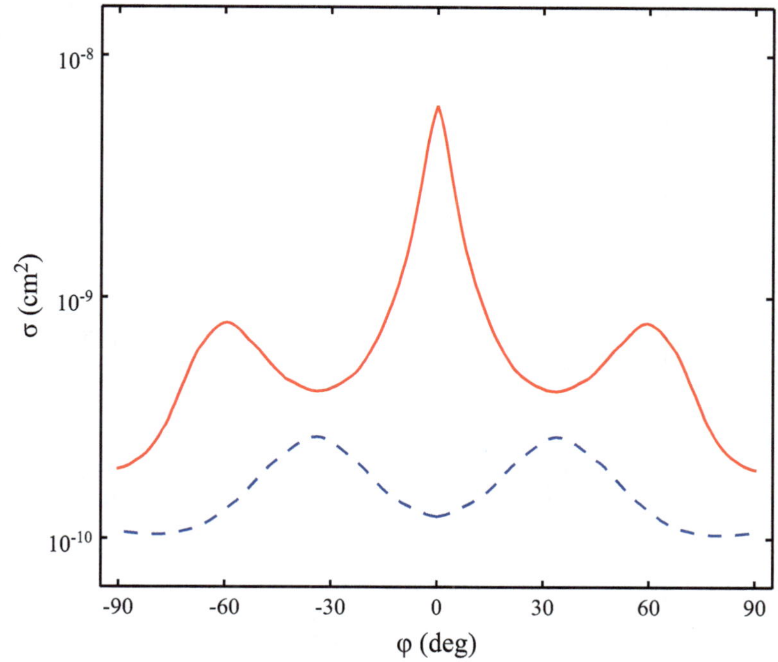


**Fig. 3.4** Dependences of the proton differential transmission cross section in the case of 7 MeV protons and the <110> channel of a 150-nm thick Si crystal on $\phi$ for $\rho = 0.085$ nm without (*dashed line*) and with (*solid line*) the effect of multiple scattering [41]

dark region is on its outer side. Since the crystal is very thin, and, hence, the majority of ions undergo one deflection from the channel wall, the rainbow we consider is the primary rainbow.

Analysis has shown that an increase of $M$ as well as introduction of the thermal vibrations of the crystal's atoms leads to a negligible change of the results presented in Figs. 3.3, 3.4, and 3.5. The latter conclusion agrees with the already mentioned fact that the rainbow line in the IP plane is "repelled" by the atomic strings.

But how should one calculate the angular distribution of ions transmitted through a very thin crystal? Such a distribution can be obtained experimentally, which would enable one to verify the model of crystal rainbows. The distribution is given by Eqs. (3.3), (3.4), (3.5), and (3.6), leading to

$$\sigma = f\left[x(\theta_x, \theta_y), y(\theta_x, \theta_y)\right] = \tilde{f}(\theta_x, \theta_y). \tag{3.12}$$

However, $x$ and $y$ are complicated multivalued functions of $\theta_x$ and $\theta_y$ [104]. Therefore, in order to avoid the mathematical difficulties, the Monte Carlo method is used. The values of $x$ and $y$ are chosen randomly from the uniform distribution within the region of the channel, and, according to the calculated values of $\theta_x$ and $\theta_y$, each ion is recorded as if it has entered a bin in the TA plane. In choosing the values of $x$ and $y$, the circular regions around the atomic strings

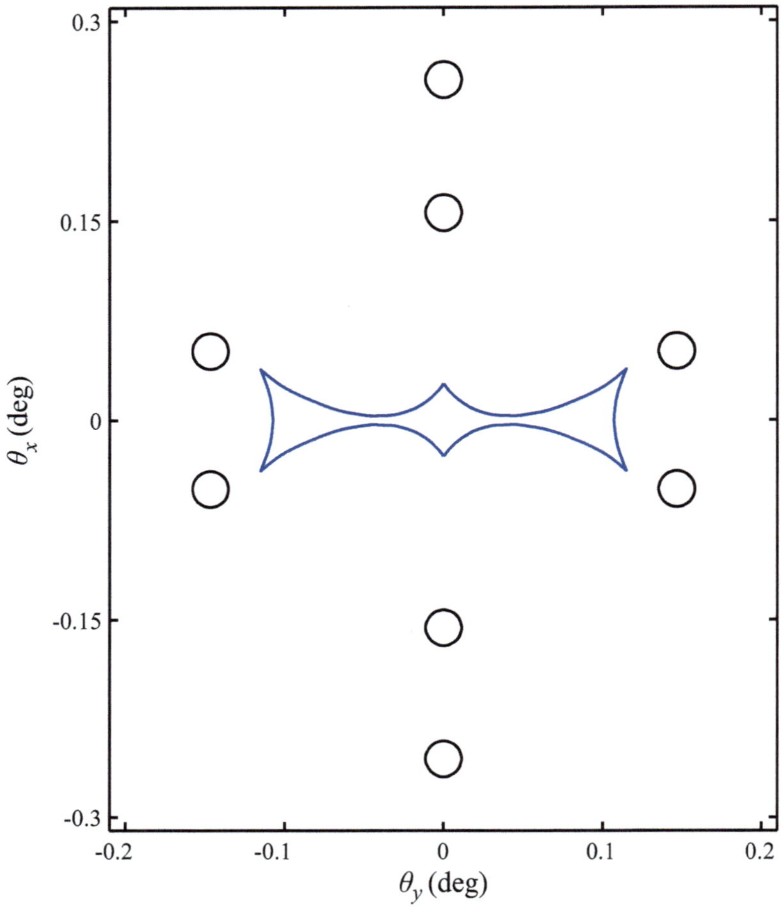


**Fig. 3.5** Rainbow line in the TA plane for 7 MeV protons transmitted through the <110> channel of a 150-nm thick Si crystal [41]. *Eight small open circles* correspond to the last atoms of the atomic strings lying on the first and second coordination circles as viewed from the channel center

having the radii equal to $a_{\mathrm{TF}}$ are avoided, and, thus, large values of the ion transmission angle are eliminated.

Now, we take into account the thermal vibrations of the crystal's atoms. They are included in the calculations as it has been explained in Subsect. 2.3.2, i.e., by averaging $V_i$ over the transverse displacements of the atoms from their equilibrium positions [78]. The atomic displacements along the $x$ and $y$ axes are taken to be small and independent, and described by a Gaussian distribution function. As a result, we obtain the thermally averaged continuum interaction potential of the ion and $i$th atomic string of the crystal,

$$V_{\mathrm{L}i}^{\mathrm{th}} = V_{\mathrm{L}i} + \frac{\sigma_{\mathrm{th}}^2}{2}(\partial_{xx}V_{\mathrm{L}i} + \partial_{yy}V_{\mathrm{L}i}), \tag{3.13}$$

where $\sigma_{th}$ is the standard deviation of the distribution, i.e., the one-dimensional atomic thermal vibration amplitude. The chosen value of $\sigma_{th}$ is 0.00744 nm, corresponding to the room temperature [108, 109]. Also, we now take into account the ion collisions with the crystal's electrons but only partly. Since the crystal is very thin, the ion energy loss caused by these collisions is disregarded. However, we include the uncertainty of the ion transmission angle appearing due to these collisions. The corresponding average square uncertainty of the ion transmission angle is

$$\sigma_e^2 = \frac{\pi Z_1 e^4 L}{E^2} n_e L_e, \tag{3.14}$$

where $n_e$ is the average density of the crystal's electrons along the straight line parallel to the channel axis determined by the average ion position in the transverse position (TP) plane, while $L_e$ is a function of the magnitude of the incident ion velocity vector, $v_0$, and $n_e$ [66]. For $L_e$, we use the expression valid in the case when $v_0$ is much greater than the maximal velocity of the crystal's electrons [110]. With the assumption that the components of the ion channeling angle are linear functions of time, the average ion position in the TP plane is determined by $x_{av} = x + (L\theta_x)/6$ and $y_{av} = y + (L\theta_y)/6$.

The angular distribution of transmitted ions with the thermal vibrations of the crystal's atoms and the uncertainty of the ion transmission angle included is generated as a sum of the two-dimensional Gaussian distribution functions located at the centers of the bins with the maximal yield of the distribution corresponding to one particular bin proportional to the number of ions that have entered the bin and its dispersion equal to the average of $\sigma_e^2$ within the bin. In the calculations, we use Eqs. (3.3) and (3.4) with $V_{Li}$, given by Eq. (3.11), substituted with $V_{Li}^{th}$, obtained by Eq. (3.13). We are giving here only the final conclusions on the obtained angular distribution. They are: (i) the rainbow line shown in Fig. 3.5 appears as a "skeleton" of the generated angular distribution of transmitted ions; (ii) the thermal vibrations of the crystal's atoms cause a negligible smearing of the angular distribution; and (iii) the ion collisions with the crystal's electrons represent a source of moderate smearing of the distribution [40, 41].

The consideration of angular distributions of ions transmitted through the axial channels of very thin crystals obtained using the model of crystal rainbows will be continued in the next subsection.

### 3.1.2 First Measurements of the Crystal Rainbows

As it has been described in the previous subsection, the crystal rainbow effect was predicted by Nešković in 1983, in the Oak Ridge National Laboratory (Tennessee, USA) [40, 41]. Soon after that, the effect was experimentally observed by a group from the same laboratory [42, 43]. That was the first quantitative study of ion channeling in thin crystals. The projectiles were 7 MeV protons, and the target was a 140-nm thick (100) Si crystal. The incident ion beam was slightly divergent – its angular full width at half-maximum (FWHM) was below 0.01°. The region of the crystal illuminated by the beam was about 0.5 mm in diameter. The transmitted ions were registered with a two-dimensional position-sensitive gas proportional counter, which included a number of vertical and horizontal anode wires and a solid-state detector behind them. Its distance from the crystal was about 5.5 m. The effective angular resolutions of the counter in the vertical and horizontal directions (FWHMs) were below $\delta_{cx} = 0.010°$ and $\delta_{cy} = 0.015°$, respectively. The crystal was mounted in a goniometer that permitted alignment of the channel axis of interest to the incident beam direction. The measurements were performed with the ions propagating along the <100> and <110> axial channels.

The measured angular distribution of protons transmitted through the <110> channels of the Si crystal is given in Fig. 3.6. In this case, $N = 520$ and $L = 198$ nm.

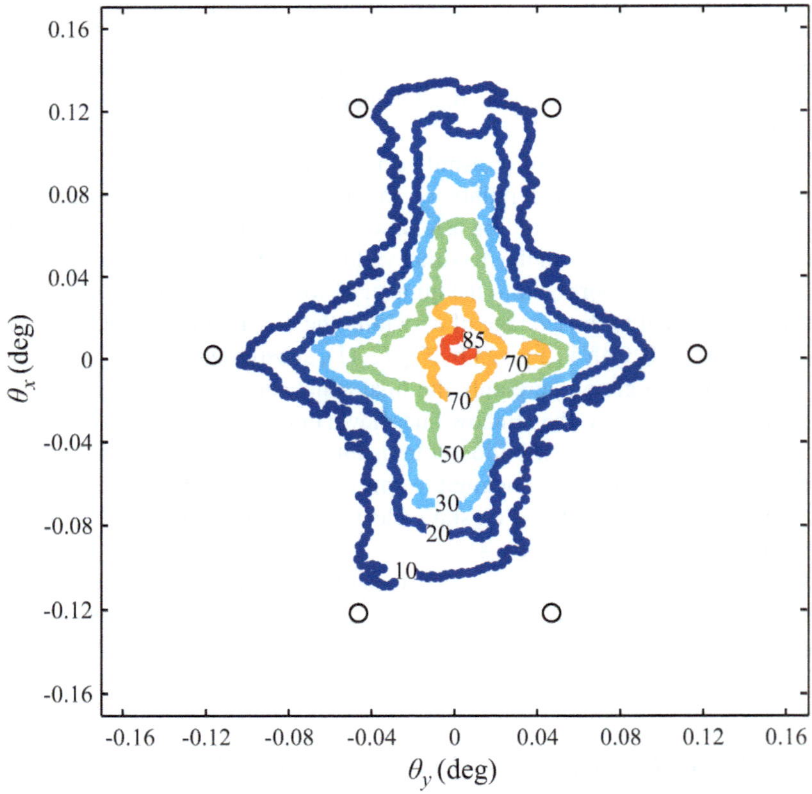


**Fig. 3.6** Experimental angular distribution of 7 MeV protons transmitted through the 198-nm long <110> channels of a Si crystal. The angular distribution is represented with six contours bounding the areas where the proton yields are above 85, 70, 50, 30, 20, and 10% of the maximal yield. *Six open circles* correspond to the last atoms of the atomic strings lying on the first coordination circle as viewed from the channel center (Adapted from Ref. [42])

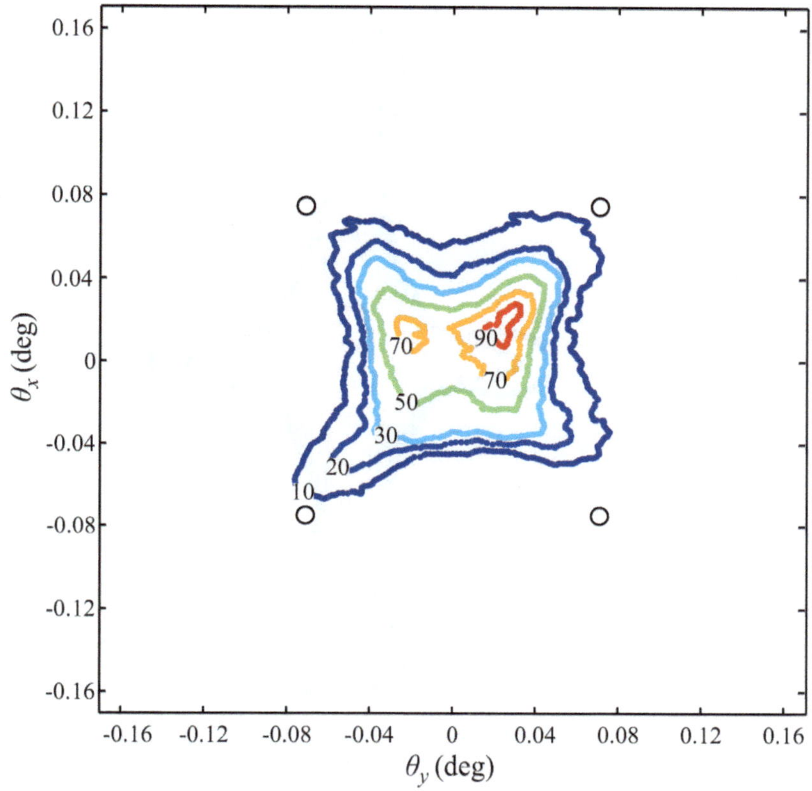


**Fig. 3.7** Experimental angular distribution of 7 MeV protons transmitted through the 140-nm long <100> channels of a Si crystal. The angular distribution is represented with six contours bounding the areas where the proton yields are above 90, 70, 50, 30, 20, and 10% of the maximal yield. *Four open circles* correspond to the last atoms of the atomic strings lying on the first coordination circle as viewed from the channel center (Adapted from Ref. [42])

This distribution is represented with six contours bounding the areas where the ion yields are above 85, 70, 50, 30, 20, and 10% of the maximal yield. Figure 3.7 gives the measured angular distribution of protons transmitted through the <100> channels of the Si crystal. In this case, $N = 260$ and $L = 140$ nm. This distribution is represented with six contours bounding the areas in which the ion yields are above 90, 70, 50, 30, 20, and 10% of the maximal yield.

Figures 3.8 and 3.9 show the calculated angular distributions corresponding to the measured distributions depicted in Figs. 3.6 and 3.7, respectively. The calculations were performed using the Monte Carlo method based on the model of crystal rainbows in the way described in the previous subsection. The sizes of a bin along the $\theta_x$ and $\theta_y$ axes were both 0.002°. For the ion-atom interaction potential, the authors used Lindhard's expression with the Thomas-Fermi atomic screening length, leading to Eqs. (3.11) and (3.13). In the case of <110> channels, they took into account the atomic strings lying on the four nearest coordination circles,

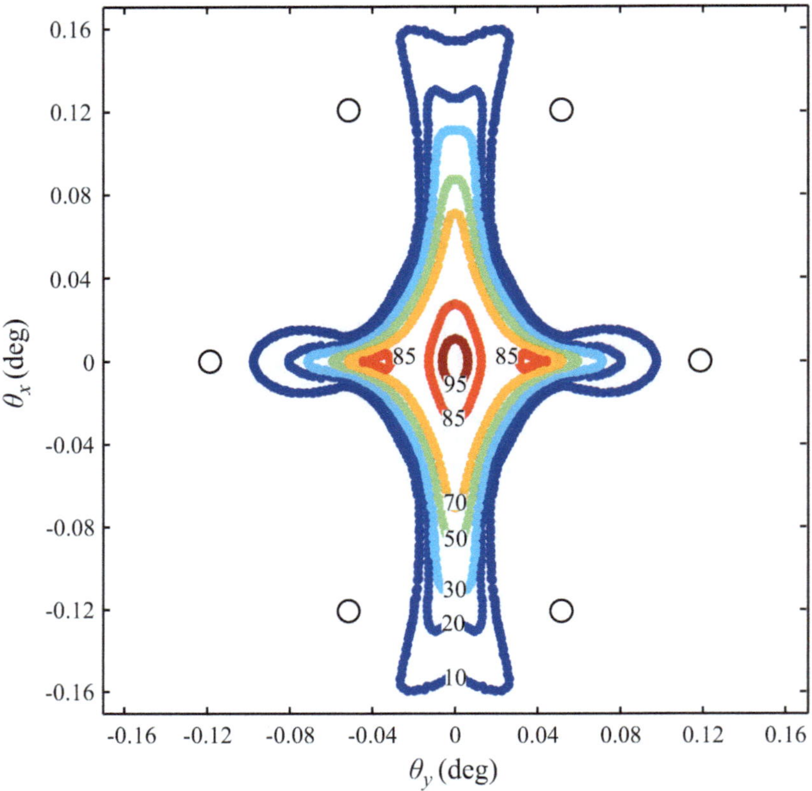


**Fig. 3.8** Theoretical angular distribution of 7 MeV protons transmitted through the 198-nm long <110> channels of a Si crystal. The angular distribution is represented with seven contours bounding the areas where the proton yields are above 95, 85, 70, 50, 30, 20, and 10% of the maximal yield. *Six open circles* correspond to the last atoms of the atomic strings lying on the first coordination circle as viewed from the channel center. The arrangement of atomic strings is rotated by 90° about the $z$ axis relative to the arrangement shown in Fig. 3.5 (Adapted from Ref. [42])

i.e., $M = 20$, while in the case of <100> channels, they included the strings lying on the three nearest coordination circles, i.e., $M = 16$. The one-dimensional atomic thermal vibration amplitude was $\sigma_{\text{th}} = 0.00744$ nm [108, 109]. The best agreement of the calculated distributions with the measured distributions was achieved when the fitting parameter was set at $C_{\text{L}} = 4.00$. Each distribution was generated as a sum of the two-dimensional Gaussian distribution functions located at the centers of the bins with the maximal yield of the distribution corresponding to one particular bin proportional to the number of ions that have entered the bin, and its vertical and horizontal FWHMs equal to $\delta_{cx}$ and $\delta_{cy}$, respectively. The authors found that, since the employed values of $\delta_{cx}$ and $\delta_{cy}$ were their maximal values, there was no need to include the uncertainty of the ion transmission angle caused by the ion collisions with the crystal's electrons.

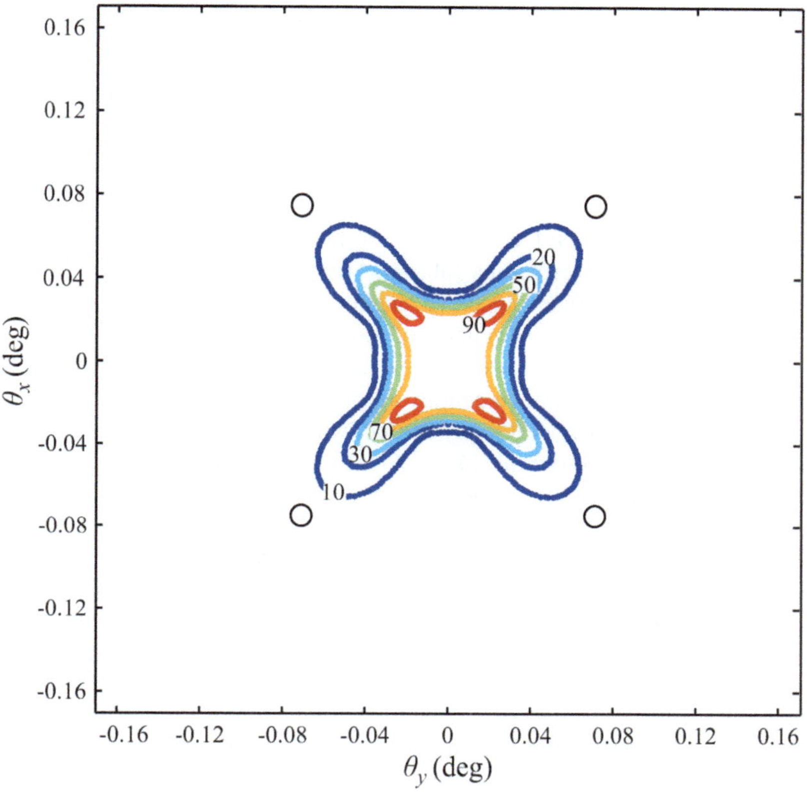


**Fig. 3.9** Theoretical angular distribution of 7 MeV protons transmitted through the 140-nm long <100> channels of a Si crystal. The angular distribution is represented with six contours bounding the areas where the proton yields are above 90, 70, 50, 30, 20, and 10% of the maximal yield. *Four open circles* correspond to the last atoms of the atomic strings lying on the first coordination circle as viewed from the channel center (Adapted from Ref. [42])

Comparison of the corresponding contours of the experimental and theoretical angular distributions of protons transmitted through the <110> channels shows that their shapes and extents are very close to each other. However, the experimental distribution contains one maximum off the origin, represented by the 70% contour, instead of two such maxima of the theoretical distribution, represented by the 85% contours. The authors attributed that discrepancy to very small tilts of the crystal about the vertical and horizontal axes, occurring due to an imperfection of the goniometer. One can see that the shapes of the 50, 30, 20, and 10% contours follow the shape of the rainbow line in the TA plane depicted in Fig. 3.5. Comparison of the corresponding contours of the measured and calculated angular distributions of protons transmitted through the <100> channels shows that their shapes and extents are very close to each other as well. In this case, the measured distribution contains two maxima off the origin, represented by the 90 and 70% contours, instead of four such maxima of the calculated distribution, represented by the 90% contours. That discrepancy was also attributed to the goniometer imperfection. As one would expect, the shapes of the 50, 30, 20, and 10% contours follow the shape of the rainbow line in the TA plane depicted in Fig. 3.16, which will be explained later.

We would like to mention that the crystal rainbow effect may have been observed but not explained before the above-described measurements of the Oak Ridge group [42, 43]. For example, Golovchenko [111] performed a transmission experiment with 2 MeV protons and a 93-nm thick (110) Si crystal. The results of the experiment were described as follows: "In the position corresponding to the initial beam direction, there appeared an intense spot with small starlike arms projecting outwards. ... The symmetry of the spots is indicative of the transverse lattice of <110> strings. ... It does not appear possible that these observations can be explained within the classical continuum picture. ... It is possible that these data represent spurious results not directly caused by the interaction of the ions with the crystal." However, Krause et al. [42] analyzed that experimental case and found that the obtained results were explainable using the model of crystal rainbows. A similar conclusion might be deduced upon analysis of the transmission experiments of Armstrong et al. [112, 113].

In continuation of its investigation of ion channeling in very thin crystals, the Oak Ridge group [44] performed a series of quantitative measurements of the angular distributions of 2–9 MeV protons and 6–30 MeV $C^{4+}$, $C^{5+}$, and $C^{6+}$ ions transmitted through the <100> channels of a 179-nm thick Si crystal. In that case, $N = 330$. The divergence of the incident ion beams (FWHM) was below 0.005°. The region of the crystal illuminated by the beam was about 0.18 mm in diameter. The transmitted ions were registered with a two-dimensional position-sensitive detector, which included a microchannel plate and a hyperbolic anode. Its distance from the crystal was 2.5 m. The effective angular resolutions of the detector in the vertical and horizontal directions (FWHMs) were $\delta_{cx} = \delta_{cy} = 0.006^{\circ}$. The crystal was mounted in a goniometer that enabled alignment of the channel axis to the incident beam direction. That was an extraordinary experimental and theoretical study.

The angular distributions of transmitted ions obtained in those expeirments were reproduced using the LAROSE computer simulation code [81, 82], mentioned in Subsect. 2.3.3. The authors used Moliére's approximation of the Thomas-Fermi ion-atom interaction potential with the Thomas-Fermi atomic screening length, defined by Eqs. (2.1), (2.2), and (2.4),

$$U_{\mathrm{M}} = \frac{Z_1 Z_2 e^2}{r_a} \sum_{\ell=1}^{3} \alpha_\ell \exp\left(-\frac{\beta_\ell r_a}{a_{\mathrm{TF}}}\right). \tag{3.15}$$

The ion energy loss in its collisions with the crystal's nuclei was disregarded [66], while the thermal vibrations of the crystal's atoms and the ion collisions with the crystal's electrons were taken into account. The agreement between the experimental and theoretical distributions was excellent – the authors were able to duplicate all the details of the distributions for all the values of the incident ion charge and kinetic energy. It must be noted that they also used in the calculations Moliére's interaction potential with Firsov's ion-atom screening length [73], $a_F$, given by Eq. (2.5), instead of $a_{TF}$. However, in those cases, they could not achieve a satisfactory agreement between the measured and calculated results. In addition, the authors discovered that the evolution of the distribution with the increase of $L$ could be precisely followed via only one variable, named the reduced crystal thickness,

$$\Lambda = \frac{f(m_1, q_1)L}{v_{av}}, \tag{3.16}$$

where $f(m_1, q_1)$ is the frequency of ion oscillations around the channel axis, being a function of $m_1$ and $q_1$, the ion mass and incident ion charge, respectively, and $v_{av}$ is the average magnitude of the ion velocity vector. In each of the analyzed cases, the value of $f$ was calculated via the value of $L$ for which the width of the distribution was minimal for the first time, corresponding to $\Lambda = 0.5$. That precious result was in fact the discovery of a scaling law in ion channeling in thin crystals, which has been confirmed in numerous later studies [91, 92, 94–100, 80, 101, 102]. If one assumes that the ion oscillations around the channel axis are harmonic, their frequency can be deduced from the second order terms of the Taylor expansion of the ion-crystal continuum interaction potential in the vicinity of the channel center. Then, it is designated as $f_h$.

Figure 3.10 shows the measured angular distribution of 30 MeV $C^{6+}$ ions transmitted through the <100> channels of a 179-nm thick Si crystal obtained by the Oak Ridge group. This distribution is represented with five contours bounding the areas in which the ion yields are above 75, 50, 20, 10, and 5% of the maximal yield. Figure 3.11 depicts the experimental yield of transmitted ions along the $\theta_y$ axis extracted from the angular distribution given in Fig. 3.10 together with the corresponding theoretical yield. It is evident that the two dependences agree excellently – each of them contains two maxima and two shoulders. The authors attributed the shoulders of the dependences to the rainbow effect. Figure 3.10 also contains the corresponding rainbow pattern in the TA plane obtained in our recent analysis performed using the theory of crystal rainbows [97] and the ion-atom interaction potential given in Sect. 4.1, which contains four small and four very small cusped isosceles triangular lines away from the origin [114]. The small rainbow lines lie on the lines $\theta_y = \pm\theta_x$, and the very small lines on the $\theta_x$ and $\theta_y$ axes. The figure clearly shows that the rainbow pattern appears as the "skeleton" of the distribution. Besides, it is evident from Figs. 3.10 and 3.11 that the two observed maxima can be connected to the two pairs of small rainbow lines lying close to the $\theta_y$ axis, and the two observed shoulders to the two very small lines lying on the $\theta_y$ axis.

Let us conclude this subsection with the more precise definitions of very thin, thin, thick, and very thick crystals than those given in the introductory part of this chapter, which are formulated via $\Lambda$. One can say that the crystal is very thin if $\Lambda$ is

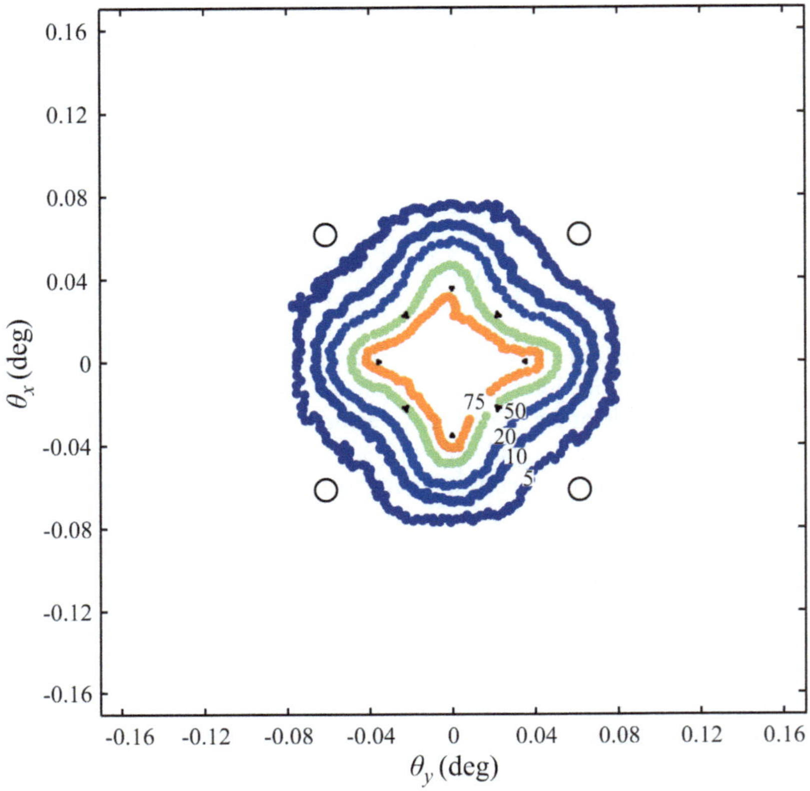


**Fig. 3.10** Experimental angular distribution of 30 MeV $C^{6+}$ ions transmitted through the 179-nm long <100> channels of a Si crystal. The angular distribution is represented with five contours bounding the areas where the ion yields are above 75, 50, 20, 10, and 5% of the maximal yield. *Four open circles* correspond to the last atoms of the atomic strings lying on the first coordination circle as viewed from the channel center. The *black lines* are the associated rainbow lines in the TA plane [114] (Adapted from Ref. [44])

smaller than about 0.25; it is thin if $\Lambda$ is between about 0.25 and about 1; it is thick if $\Lambda$ is larger than about 1; and it is very thick if $\Lambda$ is much larger than about 1 [101].

## 3.2 Crystal Rainbows as Elementary Catastrophes

### *3.2.1 Crystal Rainbow Effect as a Catastrophic Effect*

Here, we shall apply the model of crystal rainbows, described in Subsect. 3.1.1, to the transmission of protons of incident kinetic energy of $E = 10$ MeV through the <100> channel of an Au crystal of thickness of $L = 100$ nm [85–87]. The aim of the

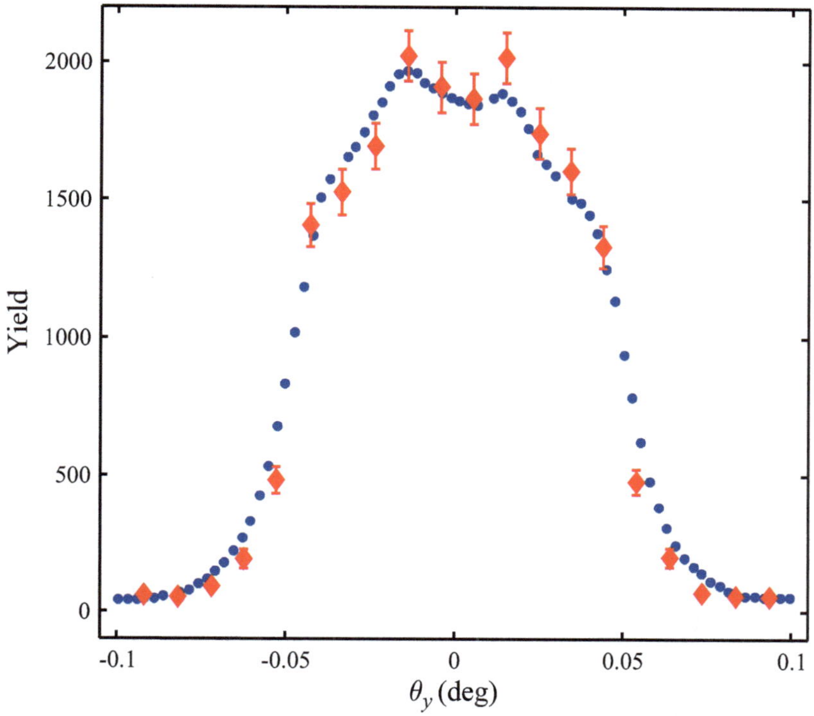


**Fig. 3.11** Experimental yield of 30 MeV $C^{6+}$ ions transmitted through the 179-nm long <100> channels of a Si crystal along the $\theta_y$ axis (*full circles*) and the corresponding theoretical yield (*full diamonds with error bars*) (see Fig. 3.10) (Adapted from Ref. [44])

study is to connect ion channeling to catastrophe theory, formulated by Thom [13]. It should be noted that a similar effect, occurring in particle scattering from surfaces, was analyzed by Berry [115].

Catastrophe theory is a general theory of models [13, 116–120]. It was founded on the principle of structural stability, and, thus, was oriented toward reality. This means that the theory does not belong to the mathematics proper. It enables one to approximate a smooth (structurally stable) function in the vicinity of its degenerate critical point with an equivalent polynomial. Such a polynomial is called the generating function of an elementary catastrophe – $G(\xi_1, \xi_2, \ldots, \xi_m; \alpha_1, \alpha_2, \ldots, \alpha_n)$, where $\xi_i$, $i = 1, 2, ..., m$, are the state variables, $m$ is the corank, $\alpha_j$, $j = 1, 2, ..., n$, are the control variables, and $n$ is the codimension of the catastrophe. The set of critical points of $G$ is called the equilibrium set, the set of its degenerate critical points the catastrophic set, and the set of projections of its critical points to the control space the bifurcation set of the catastrophe. The bifurcation set is the set of points at which a catastrophe occurs – the number of values of the mapping of the control space to the equilibrium set (discontinuously) changes.

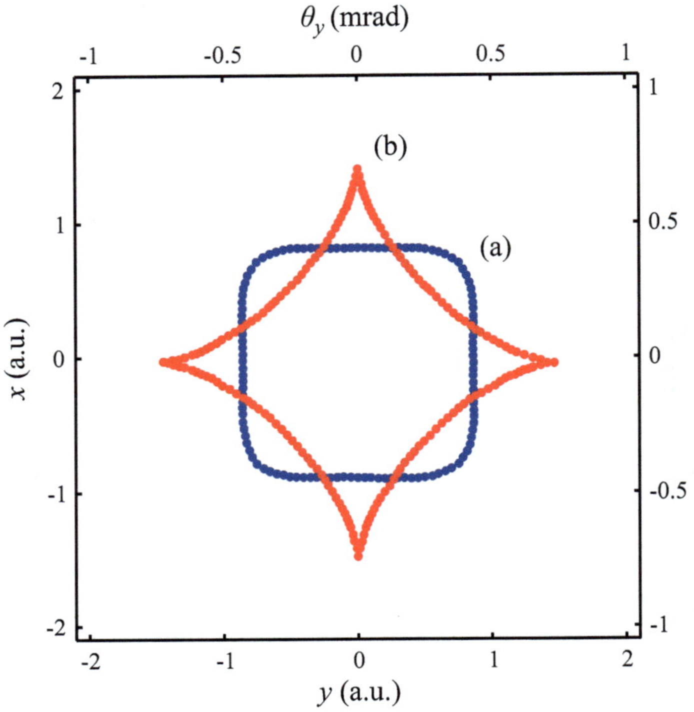


**Fig. 3.12** (**a**) Rainbow line in the IP plane for 10 MeV protons transmitted through the <100> channel of a 100-nm thick Au crystal [85]. (**b**) Associated rainbow line in the TA plane [85]. The arrangement of atomic strings is rotated by 45° about the $z$ axis relative to the arrangement used to obtain the angular distribution of transmitted ions shown in Figs. 3.7, 3.9, and 3.10

In the case under consideration, the distance between the atoms of the atomic strings of the crystal is $d = 0.407897$ nm [121]. Hence, the number of atoms in one atomic string is $N = 246$. We employ Lindhard's proton-atom interaction potential, given by Eq. (3.10), and the continuum approximation [66], which lead to Eq. (3.11), with $Z_1=1$ and $Z_2=79$. The fitting parameter is set at $C_L = 3.00$. The corresponding value of the reduced crystal thickness is $\Lambda = 0.16$, which means that the crystal is very thin. This value is calculated by Eq. (3.16) in which $v_{av}$ is taken to be the magnitude of the incident proton velocity vector and $f = f_h$, i.e., it is deduced from the second order terms of the Taylor expansions of the proton-crystal continuum interaction potential in the vicinity of the channel center. The number of atomic strings is $M = 24$, i.e., we take into account the strings lying on the four nearest coordination circles. The thermal vibrations of the crystal's atoms and the proton collisions with the crystal's electrons are neglected. Figure 3.12 gives the rainbow line in the IP plane. The transverse coordinates of the atomic strings lying on the first coordination circle, which define the channel, in atomic units (a.u.), are: $(x_i, y_i) = (2.73, 0)$, $(0, 2.73)$, $(-2.73, 0)$, and $(0, -2.73)$, with 1 a.u. $= 0.0529$ nm. As it has been said in Subsect. 3.1.1, the coordinates of the points of the rainbow line in the IP plane are determined solely by the arrangement of atomic strings and their continuum potential – they do not depend on $Z_1$, $E$, and $L$. This figure also gives the rainbow line in the TA plane, which is a cusped square. Each cusp of this line is directed toward an atomic string defining the channel. The coordinates of the points of the rainbow line in the TA plane depend on $Z_1$, $E$, and $L$. However, the ratios of these coordinates and $Z_1L/E$ do not depend on these parameters. Hence, the shape of this line is determined solely by the continuum potential in the channel. Since, as in the case analyzed in Subsect. 3.1.1, the crystal is very thin, the rainbow under investigation is the primary rainbow.

The components of the ion transmission angle, $\theta_x$ and $\theta_y$, are single-valued functions of the components of the ion impact parameter vector, $x$ and $y$. On the other hand, $x$ and $y$ are complicated multivalued functions of $\theta_x$ and $\theta_y$. One can obtain these functions by a computer simulation method. The values of $x$ and $y$ are chosen randomly or uniformly from the uniform distribution within the region of the channel and, according to the calculated values of $\theta_x$ and $\theta_y$ and chosen values of $x$ and $y$, each ion is recorded as if it has entered a bin in the $\theta_x\theta_y x$ space and a bin in the $\theta_x\theta_y y$ space. Thus, for a large number of incident ions, one obtains for one bin in the TA plane, i.e., the $\theta_x\theta_y$ space, the distributions of $x$ and $y$. The numbers of maxima of these distributions represent the numbers of values of $x$ and $y$ for this particular bin. The angular distribution of transmitted ions is generated by summing the contents of the bins in the $\theta_x\theta_y x$ or $\theta_x\theta_y y$ space corresponding to each bin in the TA plane.

Let us first explore the function $y(\theta_x, \theta_y)$ in the vicinity of the apex of the cusp of the rainbow line in the TA plane lying in its fourth and first quadrants. The change of the function $x(\theta_x, \theta_y)$ in this region, along the rainbow line, is very small (see Fig. 3.12) and, therefore, less interesting. The sizes of a bin in the $\theta_x\theta_y y$ space along the $\theta_x$ and $\theta_y$ axes both equal 0.02 mrad, and its size along the $y$ axis is 0.04 a.u. The values of $x$ and $y$ are chosen uniformly, and the number of incident ions is about 900,000. Figure 3.13 gives the changes of $y$ with $\theta_y$ for three values of $\theta_x$ – 0.51, 0.65, and 0.79 mrad. Comparison of this figure and Fig. 3.12 shows that $y$ is triple-valued on the inner side of this part of the rainbow line and single-valued on its outer side. This means that the inner side of this part of the rainbow line is the bright side of the rainbow and that its outer side is the dark side of the rainbow. Besides, it is clear that $y$ is an odd function of $\theta_y$. Analogous analyses of the function $x(\theta_x, \theta_y)$ or $y(\theta_x, \theta_y)$ in the vicinities of the apices of the other three cusps of the rainbow line have shown that the inner side of the whole rainbow line is the bright side of the rainbow and its outer side the dark side of the rainbow.

Now, we shall compare the shapes of the rainbow line, the line separating the branches of the surface defined by the function $y(\theta_x, \theta_y)$, and the surface defined by this function in the vicinity of the apex of the cusp lying in the fourth and first quadrants of the TA plane with the corresponding shapes obtained using catastrophe theory [13, 116–120]. It is clear that the problem we consider is one-dimensional. Therefore, the corank of the elementary catastrophe we want to find is one, i.e., it has one behavior variable. Analysis of the bifurcation sets of the catastrophes of this corank has shown that the first appropriate catastrophe is the

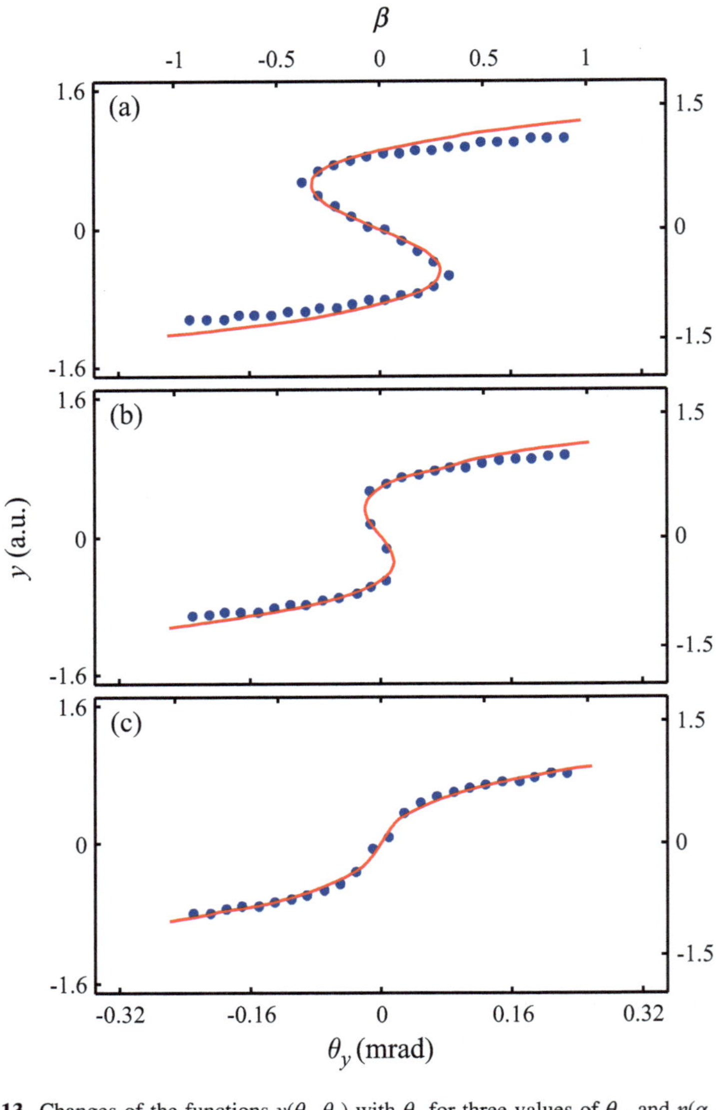


**Fig. 3.13** Changes of the functions $y(\theta_x, \theta_y)$ with $\theta_y$ for three values of $\theta_x$, and $\eta(\alpha, \beta)$ for three values of $\alpha$ in the vicinity of the apex of the cusp obtained by the computer simulation method (*full circles*) and by Eq. (3.18) (*solid lines*) for 10 MeV protons transmitted through the <100> channel of a 100-nm thick Au crystal [85]. The values of $\theta_x$ and $\alpha$ are: (**a**) 0.51 mrad and 0.886, (**b**) 0.65 mrad and 0.317, and (**c**) 0.79 mrad and −0.253, respectively

cusp catastrophe, whose codimension is two, i.e., it has two control variables. We have chosen for the behavior variable $\eta = y/\rho_0$, where $\rho_0 = k_\rho d$, $d$ is the distance between the atoms of the atomic strings, and $k_\rho$ is a spatial scaling factor. The control variables have been taken to be $\alpha = -(\theta_x - \theta_{xc})/\theta_0$ and $\beta = \theta_y/\theta_0$, where $\theta_{xc}$ is the value of $\theta_x$ corresponding to the apex of the cusp, $\theta_0 = k_\theta Z_1 Z_2 e^2$

$N/(2E)$, and $k_\theta$ is an angular scaling factor. For the generating function of the cusp catastrophe, we take its dual form,

$$G_2 = -\eta^4 + 2\alpha\eta^2 + 4\beta\eta. \quad (3.17)$$

One can easily find that the equilibirum set of this catastrophe is the surface defined by

$$\eta^3 - \alpha\eta - \beta = 0, \quad (3.18)$$

which corresponds to the surface defined by $y(\theta_x, \theta_y)$, that its catastrophic set is the line defined by

$$\eta = \pm\left(\frac{\alpha}{3}\right)^{1/2}, \quad (3.19)$$

which separates the branches of the surface defined by $y(\theta_x, \theta_y)$, and that its bifurcation set is the line determined by

$$\alpha = 3\left|\frac{\beta}{2}\right|^{2/3}, \quad (3.20)$$

which corresponds to the rainbow line in the vicinity of the apex of the cusp.

Figure 3.14 shows the comparison of the shapes of the rainbow line in the TA plane in the vicinity of the apex of the cusp and the line defined by Eq. (3.20).

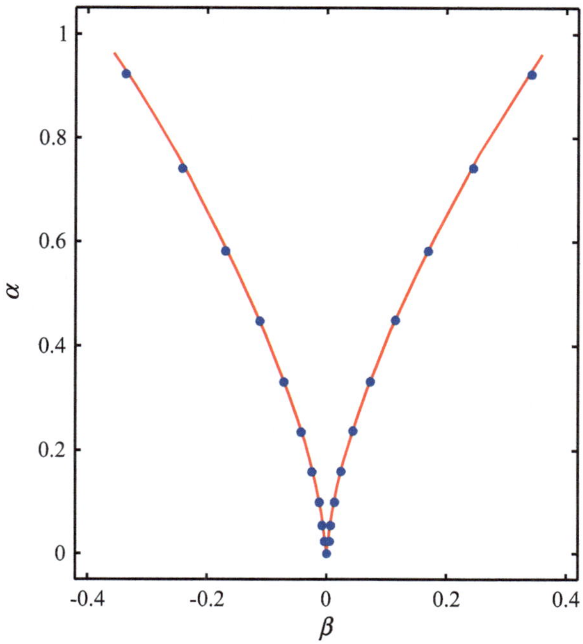


**Fig. 3.14** Rainbow line in the TA plane in the vicinity of the apex of the cusp (*full circles*) and the line defined by Eq. (3.20) (*solid line*) for 10 MeV protons transmitted through the <100> channel of a 100-nm thick Au crystal [85]

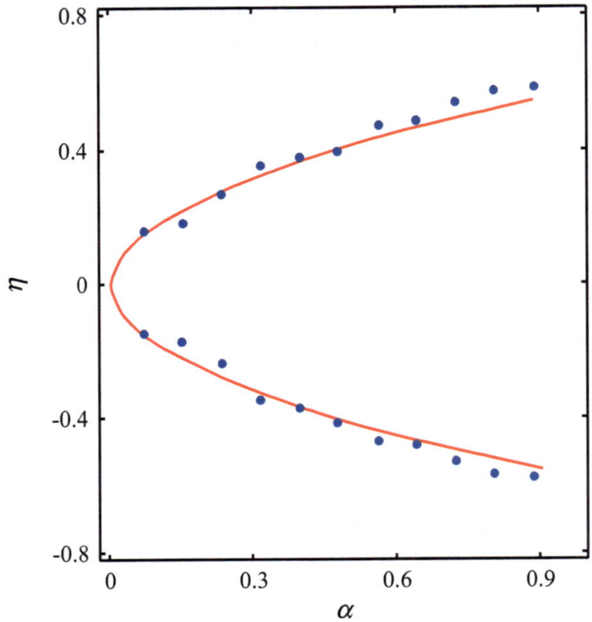


**Fig. 3.15** Lines separating the branches of the surfaces defined by the functions $\eta(\alpha, \beta)$ in the vicinity of the apex of the cusp obtained by the computer simulation method (*full circles*) and by Eq. (3.19) (*solid line*) for 10 MeV protons transmitted through the <100> channel of a 100-nm thick Au crystal [85]

Each branch of the rainbow line is given by 12 points, corresponding to the points in the IP plane with the $y$ coordinates between 0 and $\pm 0.55$ a.u. with the step of 0.05 a.u. The last point of each branch of the rainbow line is chosen so that the ratio of its $\alpha$ coordinate and $\alpha_m$, being the $\alpha$ coordinate of the midpoint of the rainbow line between the cusp under consideration and the neighboring cusp, is smaller than and close to 0.5. The value of $k_\theta$ is 0.0171, which is obtained via the condition that the average ratio of the $\alpha$ coordinates of the points of the rainbow line and the corresponding points of the line given by Eq. (3.20) equals 1. Comparison of the shapes of the lines separating the branches of the surfaces determined by the functions $\eta(\alpha, \beta)$ obtained by the computer simulation method and Eq. (3.19) is depicted in Fig. 3.15. Each branch of the former line is given by 11 points, whose $\alpha$ coordinates are determined by the $\theta_x$ coordinates between 0.51 and 0.71 mrad, with the step of 0.02 mrad. As in case of the rainbow line, the last point of each branch of this line is chosen so that the ratio of its $\alpha$ coordinate and $\alpha_m$ is smaller and close to 0.5. The value of $k_\rho$, which is obtained in the analogous way as the value of $k_\theta$, is 0.120. The shapes of the surfaces under consideration are compared in Fig. 3.13. The corresponding values of $\alpha$ are 0.886, 0.317, and $-0.253$, respectively.

One can conclude that the agreement between the shapes of the rainbow lines in the TA plane, of the lines separating the branches of the surfaces determined by the functions $\eta(\alpha, \beta)$, and of the surfaces defined by these functions in the vicinity of the apex of the cusp calculated by the model of crystal rainbows and catastrophe theory is excellent. Thus, we have demonstrated that the crystal rainbow effect can be accurately modeled using catastrophe theory. One can say that the effect is a catastrophic effect.

### 3.2.2 The $^4X_9$ Catastrophe as the Organizing Center of Crystal Rainbows

In the previous subsection, we have presented an accurate model of the rainbow effect occurring in ion transmission through a channel of a very thin crystal created using catastrophe theory. However, the modeling has been performed only in the vicinity of a cusp of the rainbow line in the TA plane. But is there a catastrophic model that will comprise the whole rainbow line appearing for this and other arrangements of atomic strings? In order to answer this question, we are going to apply the model of crystal rainbows to the cases of square, rectangular, centered rectangular, and hexagonal very thin crystals with one atomic string per channel [88–90, 93]. These are the cases of all four special two-dimensional Bravais lattices with one atomic string per lattice point.

The projectiles will be 10 MeV protons, and the targets 100 nm thick <100>, <110>, and <111> Au and W crystals[1] [88]. These crystals are very thin, i.e., $\Lambda < 0.25$. It is well known that an Au crystal has a face-centered cubic and a W crystal a body-centered cubic structure. Consequently, the <100> Au or W crystal is a square, the <110> W crystal a rectangular, the <110> Au crystal a centered rectangular, and the <111> Au or W crystal a hexagonal very thin crystal with one atomic string per channel. We use Lindhard's ion-atom interaction potential, leading to Eq. (3.11). Again, $C_{\mathrm{L}} = 3.00$ and $M = 24$. The thermal vibrations of the crystal's atoms and the ion collisions with the crystal's electrons are not taken into account. It is easy to see that $\theta_x$ and $\theta_y$, determined by Eqs. (3.3) and (3.4), can be also obtained from the function

$$F = x\theta_x + y\theta_y + \frac{L}{2E}(V - V_0), \tag{3.21}$$

with

$$V = \sum_{i=1}^{M} V_i, \tag{3.22}$$

where $V_i$ is the continuum interaction potential of the ion and $i$th atomic string, $V$ is the ion-crystal continuum interaction potential, and $V_0$ is the value of $V$ at the channel center, via the conditions $\partial_x F = 0$ and $\partial_y F = 0$. Variable $V/(Z_1 e)$ is the continuum potential in the channel. The function $F$ is the reduced action function of the ion [8], while the two above-introduced conditions represent the principle of least action in the case in question. Now, Eq. (3.6), defining the ion differential transmission cross-section, given by Eq. (3.5), can be written as

[1] An <hkl> Au or W crystal is an Au or W crystal oriented in a way to make the axes of its <hkl> channels parallel to the incident proton beam direction.

$$J = \partial_{xx} V \partial_{yy} V - \left(\partial_{xy} V\right)^2 \tag{3.23}$$

or

$$J = \partial_{xx} F \partial_{yy} F - \left(\partial_{xy} F\right)^2. \tag{3.24}$$

Thus, in addition to being the Jacobian of $\theta_x$ and $\theta_y$, $J$ is the Hessian of $V$ as well as of $F$. Equation (3.23) means that the rainbow line in the IP plane, which is defined by the equation $J = 0$, is the line along which the Gauss curvature of the continuum potential in the channel vanishes [122].

Let us now model the crystal rainbow effect by catastrophe theory. This will be done by approximating the ratio

$$\frac{F}{\rho_0 \theta_0} = \alpha\xi + \beta\eta + \frac{d(V - V_0)}{Z_1 Z_2 e^2}, \tag{3.25}$$

where $\xi = x/\rho_0$, $\eta = y/\rho_0$, $\alpha = \theta_x/\theta_0$, $\beta = \theta_y/\theta_0$, and $\theta_0 = Z_1 Z_2 e^2 N/(2E)$, with the generating function of an elementary catastrophe [8]; $\rho_0$ and $\theta_0$ are the spatial and angular scaling factors, respectively. One should note that $\alpha$ and $\theta_0$ are chosen here differently than in the previous subsection and that $\rho_0$ is left free. But how to find the appropriate catastrophe? Equation (3.25) shows that the problem we investigate is two-dimensional. This means that the corank of the catastrophe we are looking for is two, i.e., it has two state variables. Our intention to describe all four special two-dimensional Bravais lattices leads to a conclusion that our catastrophe should be nonsimple, i.e., it should be a family of catastrophes. Analysis has shown that the first family of catastrophes of this corank is the $X_9$ catastrophe, whose codimension is eight – it has one modulus and seven control variables [117]. This family has three subfamilies, which are denoted by ${}^0X_9$, ${}^2X_9$, and ${}^4X_9$. Comparison of the shapes of the bifurcation sets of all these catastrophes and the rainbow lines in the TA plane in the above-specified ion-crystal cases has indicated that the appropriate catastrophe is the ${}^4X_9$ catastrophe [123, 12, 18]. The generating function of this catastrophe is

$$G_K = \pm(\xi^4 + K\xi^2\eta^2 + \eta^4) + \gamma\xi^2 + \delta\eta^2 + \alpha\xi + \beta\eta, \tag{3.26}$$

where $\xi$ and $\eta$ are its behavior variables, $K < -2$ is its modulus, and $\alpha, \beta, \gamma$, and $\delta$ are its control variables; $\xi$, $\eta$, $\alpha$, and $\beta$ have been defined above while $K$, $\gamma$, and $\delta$ are left free; the plus and minus signs in front of the parentheses correspond to its standard and dual forms, respectively. Hence, according to Eq. (3.25), $V$ is approximated by

$$V_K = V_0 + \frac{Z_1 Z_2 e^2}{d}\left[\gamma\xi^2 + \delta\eta^2 \pm (\xi^4 + K\xi^2\eta^2 + \eta^4)\right], \tag{3.27}$$

being the catastrophic ion-crystal continuum interaction potential. This expression is invariant under the symmetry operations relative to the channel center from the

$C_{2v}$ point group, which correspond to the square, rectangular, centered rectangular, and hexagonal very thin crystals with one atomic string per channel [124]. For $\gamma = \delta$, $V_K$ is invariant under the symmetry operations relative to the channel center from the $C_{4v}$ point group, which correspond to the square very thin crystals with one atomic string per channel [124]. Therefore, one can assume that the ${}^4X_9$ catastrophe is the organizing center of crystal rainbows produced with ions and square, rectangular, centered rectangular, and hexagonal very thin crystals with one atomic string per channel.

Figure 3.16 shows the rainbow lines in the TA plane in the case of <100> Au crystal generated by the functions $F$ and $G_K$; the required form of $G_K$ is its dual form. The chosen arrangement of atomic strings defining the channel is depicted in the figure. The latter rainbow line is obtained by a fitting procedure in which point 1′ coincides with 1 and the distance of 2′ from 2 is reduced to zero by

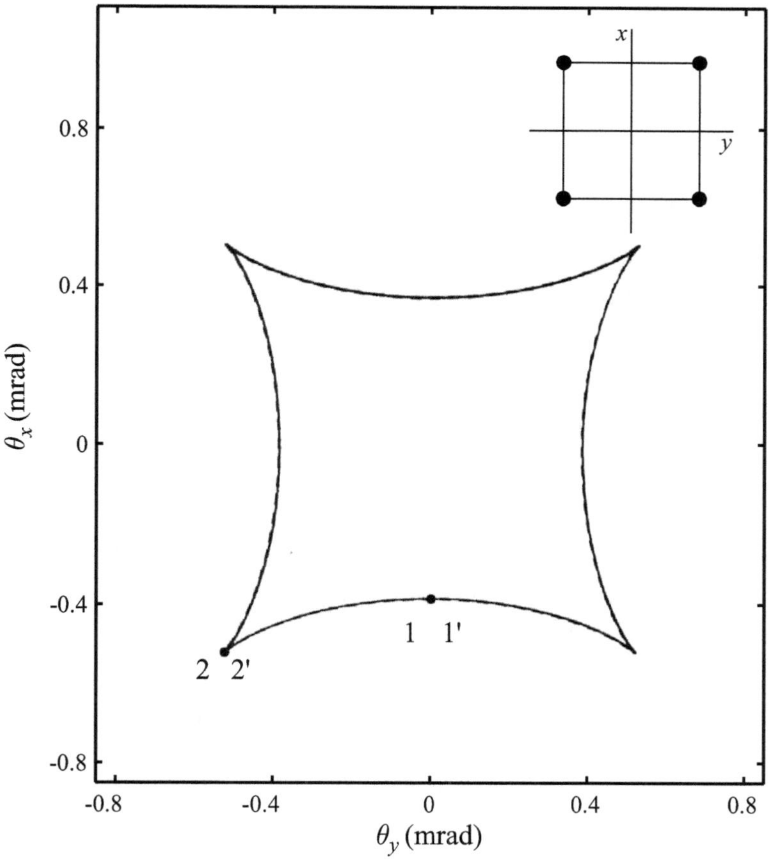


**Fig. 3.16** Rainbow line in the TA plane for 10 MeV protons and a 100-nm thick <100> Au crystal obtained by the function $F$ (*solid line*) and the function $G_K$ (*dashed line*) [88]. Points 1 and 2 belong to the former line, and points 1’ and 2’ to the latter line. The arrangement of atomic strings defining the channel is shown too

**Table 3.1** Values of the modulus, free control variables, and spatial scaling factor of the generating function of the $^4X_9$ elementary catastrophe for 10 MeV protons transmitted through the 100 nm thick <100>, <110>, and <111> Au and W crystals

| Very thin crystal | Form of $G_K$ | $K$ | $\gamma$ | $\delta$ | $\rho_0$ (a.u.) |
|---|---|---|---|---|---|
| <100> Au | Dual | −11.30 | 0.263 | 0.263 | 5.05 |
| <100> Au | Standard | −3.72 | 0.170 | 0.170 | 4.02 |
| <100> W | Dual | −11.32 | 0.245 | 0.245 | 5.74 |
| <100> W | Standard | −3.72 | 0.159 | 0.159 | 4.56 |
| <110> W | Dual | −7.45 | 0.504 | 0.160 | 4.55 |
| <110> Au | Standard | −3.69 | −0.022 | 0.314 | 5.67 |
| <111> W | Standard | −2.72 | −0.100 | 0.287 | 6.02 |
| <111> Au | Standard | −2.71 | −0.149 | 0.427 | 3.18 |

changing $K$. The resulting values of $K$, $\gamma = \delta$, and $\rho_0$ are given in Table 3.1. The same fitting procedure is applied to the case of <100> W crystal with the same arrangement of atomic strings. The obtained results are also given in Table 3.1. We have also analyzed the cases of <100> Au and W crystals with the arrangement of atomic strings rotated by 45° about the $z$ axis relative to the one indicated in the figure. In these cases, the required form of $G_K$ is its standard form. The analogous fitting procedure has led to the values of the parameters of $V_K$ given in Table 3.1. It can be shown that the values of $K$ deduced for the two arrangements of atomic strings, corresponding to the dual and standard forms of $G_K$, $K_d$, and $K_s$, respectively, are connected by the expression $K_d = -2(K_s - 6)/(K_s + 2)$ [123].

The rainbow lines in the TA plane in the case of <110> W crystal generated by the functions $F$ and $G_K$ are given in Fig. 3.17; the required form of $G_K$ is its dual form. The chosen arrangement of atomic strings defining the channel is depicted in the figure. The latter rainbow line is obtained by a fitting procedure in which point $1'$ coincides with 1 and the distance of $2'$ from 2 is minimized by changing $K$. The resulting values of $K$, $\gamma$, $\delta$, and $\rho_0$ are given in Table 3.1.

Figure 3.18 shows the rainbow lines in the TA plane in the case of <110> Au crystal generated by the functions $F$ and $G_K$; the required form of $G_K$ is its standard form. The chosen arrangement of atomic strings defining the channel is given in the figure. In this case, the channel comprises two isosceles triangular subchannels, whose centers lie on the $x$ axis above and below the channel center. The height of the wall between the upper and lower subchannels is smaller than the height of the four channel walls. Each rainbow line consists of two parts, corresponding to the two subchannels. The latter line is obtained by a fitting procedure analogous to the one used in the <110> W case. The resulting values of the parameters of $V_K$ are given in Table 3.1.

The rainbow lines in the TA plane in the case of <111> W crystal generated by the functions $F$ and $G_K$ are depicted in Fig. 3.19; the required form of $G_K$ is its standard form. The chosen arrangement of atomic strings defining the channel is shown in the figure. In this case, the channel comprises two equilateral triangular subchannels, whose centers lie on the $x$ axis above and below the channel center.

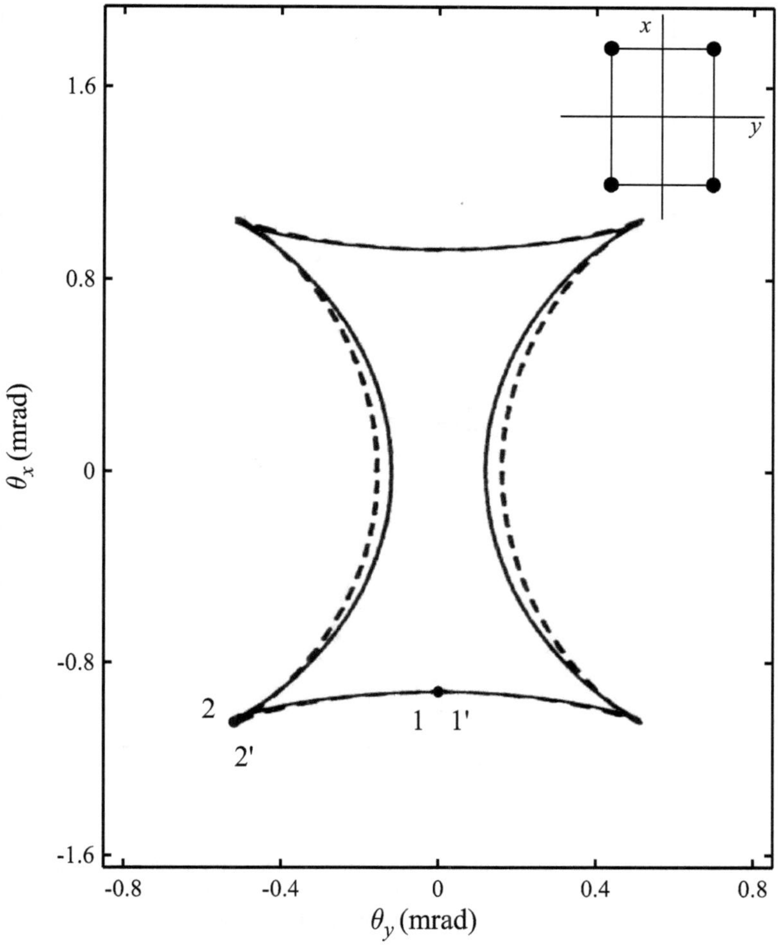


**Fig. 3.17** Rainbow line in the TA plane for 10 MeV protons and a 100-nm thick <110> W crystal obtained by the function $F$ (*solid line*) and the function $G_K$ (*dashed line*) [88]. Points 1 and 2 belong to the former line, and points 1′ and 2′ to the latter line. The arrangement of atomic strings defining the channel is shown too

The height of the wall between the upper and lower subchannels equals the height of the four channel walls. This means that the two subchannels can be also treated as two separate channels. As in the case of <110> Au crystal, each rainbow line consists of two parts, corresponding to the two subchannels. The latter line is obtained by the same fitting procedure as in the <110> W case. The resulting values of $K$, $\gamma$, $\delta$, and $\rho_0$ are given in Table 3.1. This fitting procedure has been also applied to the case of <111> Au crystal with the arrangement of atomic strings as in the case of <111> W crystal. The obtained results are given in Table 3.1 as well.

Let us now demonstrate the accurateness of the modeling of crystal rainbows using catastrophe theory. We shall take the above-treated case of <100> Au crystal

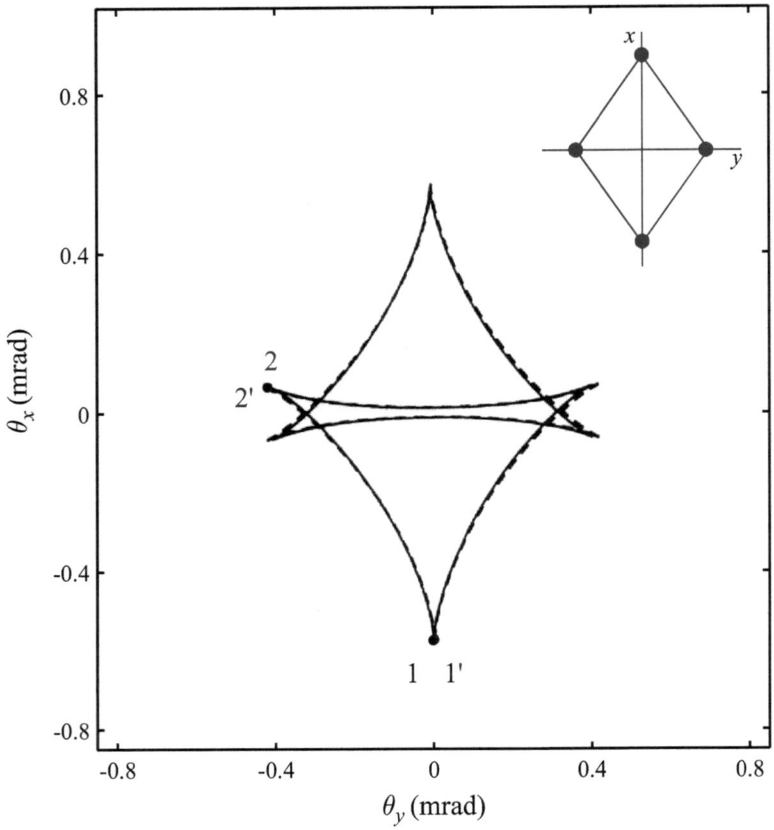


**Fig. 3.18** Rainbow line in the TA plane for 10 MeV protons and a 100-nm thick <110> Au crystal obtained by the function $F$ (*solid line*) and the function $G_K$ (*dashed line*) [88]. Points 1 and 2 belong to the former line, and points 1′ and 2′ to the latter line. The arrangement of atomic strings defining the channel is shown too

and compare the ratios $V_K/V$ and $V_T/V$, where $V_T$ is the Taylor expansion of $V$ in the vicinity of the channel center up to the fourth order terms, as functions of the variable $\rho = (x^2 + y^2)^{1/2}$ for $\phi = \tan^{-1}(y/x) = 0$. These dependences are given in Fig. 3.20. As can be expected, in the vicinity of the point $\rho$ = 1.06 a.u., which belongs to the rainbow line in the IP plane, $V_K$ is closer to $V$ than $V_T$. Besides, $V_K/V$ is between 0.99 and 1.01, i.e., $V_K$ is a good approximation of $V$, in the whole region between the channel center and wall, i.e., for $\rho$ between 0 and 1.92 a.u., while $V_T/V$ is between these two values for $\rho$ between 0 and 1.58 a.u. Comparison of these dependences for the other values of $\phi$ and the analogous results obtained in the other above-presented cases of very thin crystals has shown that $V_K/(Z_1e)$ can be used as a good approximation of $V/(Z_1e)$ in the vicinity of the rainbow line in the IP plane, and that it is a good approximation of $V/(Z_1e)$ in a larger region of the channel than $V_T/(Z_1e)$. Therefore, we may conclude that $V_K/(Z_1e)$ is a simple, universal, and

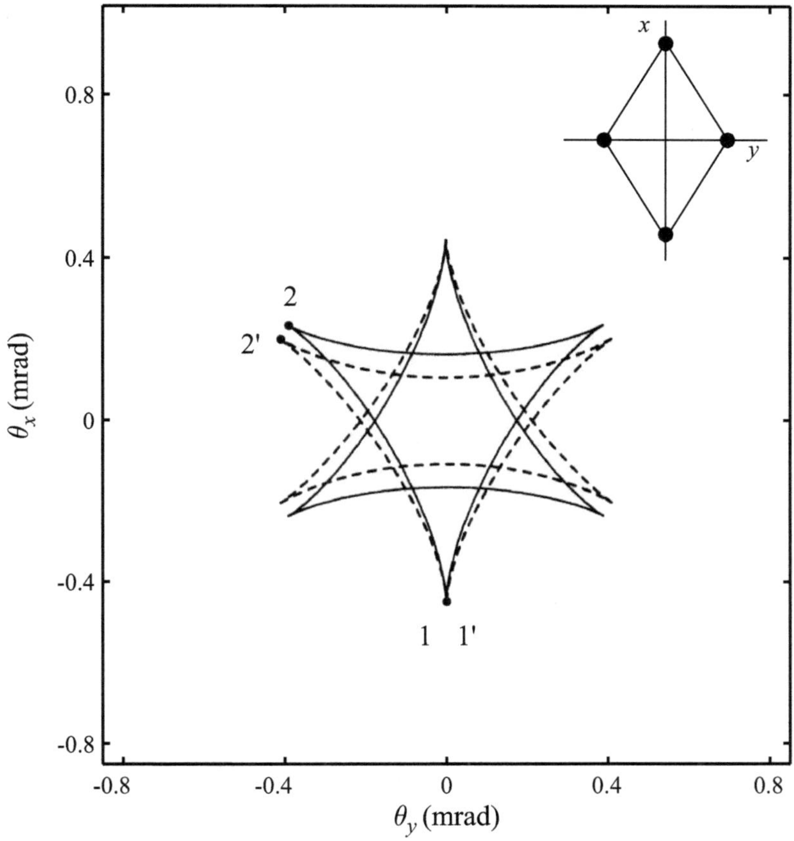


**Fig. 3.19** Rainbow line in the TA plane for 10 MeV protons and a 100-nm thick <111> W crystal obtained by the function $F$ (*solid line*) and the function $G_K$ (*dashed line*) [88]. Points 1 and 2 belong to the former line, and points 1′ and 2′ to the latter line. The arrangement of atomic strings defining the channel is shown too

relatively accurate model potential that can be employed in solving various problems in the field of ion channeling. It has been deduced starting from Lindhard's ion–atom interaction potential as the exact one. If that were done starting from a more accurate ion–atom interaction potential, e.g., the one presented in Sect. 4.1, the resulting continuum potential in the channel would be more absolutely accurate.

One can see in Table 3.1 that the values of $K$ obtained in the cases of <100> Au and W crystals with the dual form of $G_K$ as well as with its standard form, and in the cases of <111> W and Au crystals coincide or are very close to each other. This is an indication that all the square very thin crystals with one atomic string per channel can be characterized by one value of $K$, that the same is true for the hexagonal very thin crystals with one atomic string per channel, and that the same might be true for the rectangular and centered rectangular very thin crystals with one atomic string

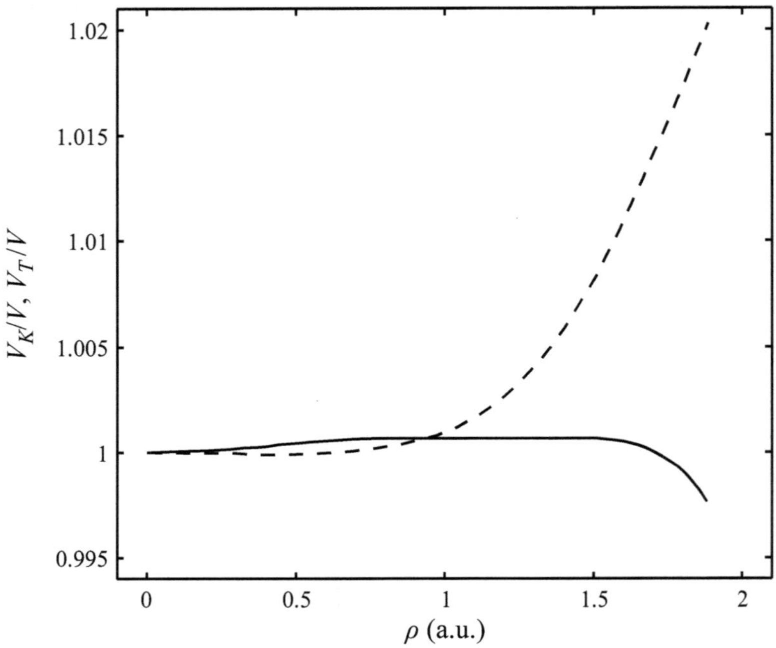


**Fig. 3.20** Ratios $V_K/V$ (*solid line*) and $V_T/V$ (*dashed line*) as functions of $\rho$ for $\phi = 0$ obtained with 10 MeV protons and a 100-nm thick <100> Au crystal [88]

per channel. In order to check this indication, we shall first investigate the other cases of square very thin crystals with one atomic string per channel.

As it has been chosen above, the projectiles will be 10 MeV protons [89]. The targets will be 100-nm thick <100> very thin crystals of eight chemical elements with the body-centered cubic structure – V, Cr, Fe, Nb, Mo, Ba, Eu, and Ta; fourteen elements with the face-centered cubic structure – Al, Ca, Ni, Cu, Sr, Rh, Pd, Ag, Ce, Yb, Ir, Pt, Pb, and Th; and four elements with the diamond-type cubic structure – C, Si, Ge, and Sn. Together with the above analyzed cases of <100> W and Au crystals, the total number of analyzed cases of square very thin crystals with one atomic string per channel will be 28. The arrangement of atomic strings will be rotated by 45° about the $z$ axis relative to the one indicated in Fig. 3.16. Consequently, the required form of $G_K$ will be its standard form. Figures 3.21 and 3.22 give the dependences of the parameters of $V_K$ on $Z_2$. It is evident that the line representing $K$ as a function of $Z_2$ is very close to a straight line parallel to the abscissa, indicating that this parameter does not depend on the position of the element in the periodic table of elements. Its average value is $K^{sq} = -3.73$, with the value of the standard deviation being 0.03. Hence, we conclude that the value of the modulus of the elementary catastrophe may be determined by the symmetry of the very thin crystal in question. One can also see that the dependences of $\gamma$ and $1/\rho_0$ on $Z_2$ are periodic, indicating that these parameters are strongly correlated with the position of the element in the periodic table. The three maxima clearly visible in the

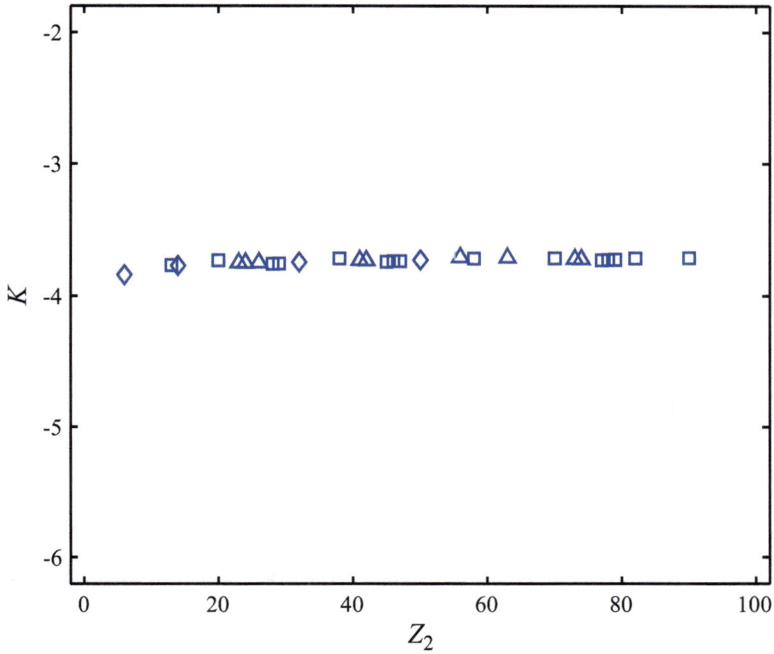


**Fig. 3.21** Dependence of the parameter $K$ of the catastrophic ion-crystal continuum interaction potential on $Z_2$ in the cases of square very thin crystals with one atomic string per channel; *triangles*, *squares*, and *diamonds* correspond to the crystals with the body-centered, face-centered, and diamond-type cubic structures, respectively [89]

dependences shown in Fig. 3.22 correspond to Ni, Rh, and Ir. The fact that the crystal density as a function of the position of the element in the periodic table is also maximal for these three elements [125] has led us to analyze the (surface) density of the <100> atomic strings, $n_\rho$, as a function of $Z_2$ for the 28 elements under consideration. This dependence is shown in Fig. 3.23. Its shape is very similar to the shapes of the dependences given in Fig. 3.22, with the three clearly visible maxima corresponding to Ni, Rh, and Ir too. Hence, we conclude that the values of the free control variables of the catastrophe and the spatial scaling factor may be proportional to the value of $n_\rho$ of the very thin crystal under consideration.

Besides, we have analyzed the crystal rainbow effect in the transmission of 10 MeV protons through 100-nm thick <110> very thin crystals of fourteen chemical elements with the face-centered cubic structure – Al, Ca, Ni, Cu, Sr, Rh, Pd, Ag, Ce, Yb, Ir, Pt, Pb and Th [93]. These are the cases of centered rectangular very thin crystals with one atomic string per channel. Together with the above-analyzed case of <110> Au crystal, the total number of such cases we have analyzed is 15. The values of the parameters of $V_K$ for different values of $Z_2$ have been obtained by a fitting procedure slightly improved relative to the one used in the case of <110> Au crystal. Again, the line representing $K$ as a function of $Z_2$

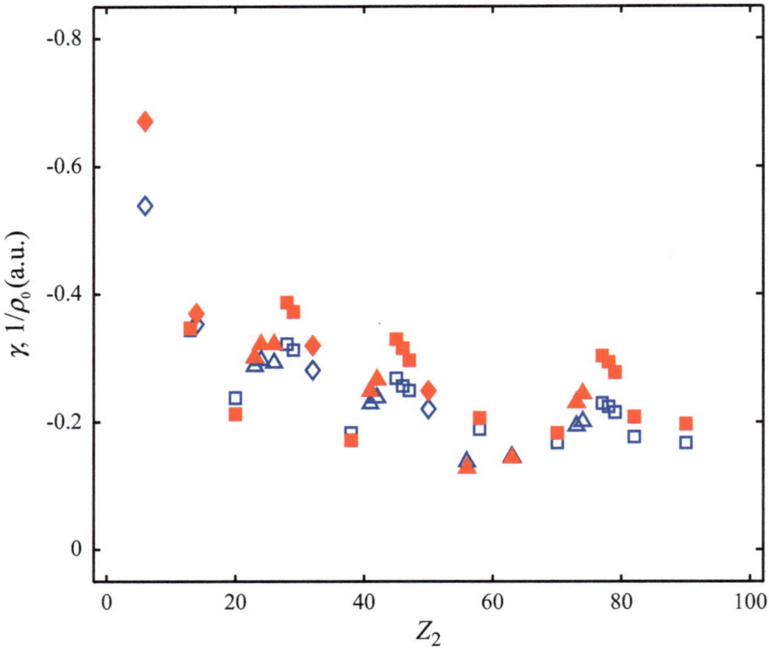


**Fig. 3.22** Dependences of the parameters $\gamma$ (*open symbols*) and $1/\rho_0$ (*full symbols*) of the catastrophic ion-crystal continuum interaction potential on $Z_2$ in the cases of square very thin crystals with one atomic string per channel; *triangles*, *squares*, and *diamonds* correspond to the crystals with the body-centered, face-centered, and diamond-type cubic structures, respectively [89]

is very close to a straight line parallel to the abscissa. The average value of this parameter is $K^{cr} = -3.81$, with the value of the standard deviation being 0.02. Thus, one can infer that the value of the modulus of the elementary catastrophe is determined by the symmetry of the very thin crystal in question. In this case, the shapes of $-\gamma$ and $\delta$ as functions of $Z_2$ are very similar to the shape of the density of the <110> atomic strings, $n_\rho$, as a function of $Z_2$. Therefore, one can conclude that the values of the free control variables of the catastrophe are proportional to the value of $n_\rho$ of the very thin crystal under consideration. This means that we have found a consistent interpretation of all the parameters of the above-introduced catastrophic ion–crystal continuum interaction potential. We consider this a clear demonstration of the fact mentioned in the previous subsection that catastrophe theory is oriented toward reality.

In addition, we shall present here the results of the analysis of the crystal rainbows produced with 10 MeV protons and 100-nm thick <111> very thin crystals of the above-chosen 28 chemical elements [90]. These are the cases of hexagonal very thin crystals with one atomic string per channel. However, unlike in the above described cases of <111> W and Au crystals, we have treated the two

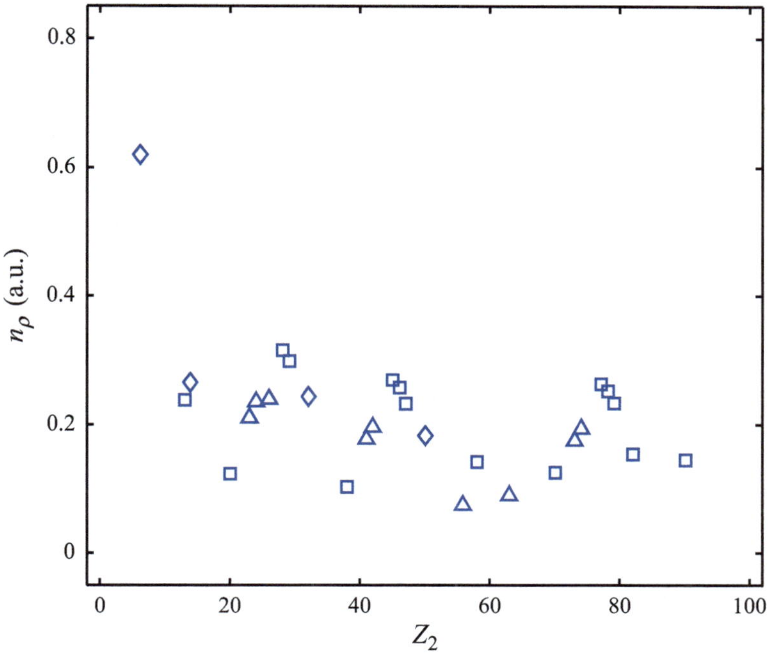


**Fig. 3.23** Dependences of the density of the <100> atomic strings on $Z_2$ in the cases of square very thin crystals with one atomic string per channel; *triangles*, *squares*, and *diamonds* correspond to the crystals with the body-centered, face-centered, and diamond-type cubic structures, respectively [89]

(equilateral triangular) subchannels of the channel as two separate channels. Analysis has shown that in this case, the appropriate elementary catastrophe is the elliptic umbilic catastrophe, which is a bordering catastrophe of the $^4X_9$ catastrophe [117]. Its corank is two, and its codimension is three. The generating function of this catastrophe is

$$G_5 = \xi^3 - 3\xi\eta^2 + \gamma(\xi^2 + \eta^2) + \alpha\xi + \beta\eta, \tag{3.28}$$

where $\xi$ and $\eta$ are its behavior variables, and $\alpha$, $\beta$, and $\gamma$ its control variables; $\xi$, $\eta$, $\alpha$, and $\beta$ have been defined above while $\gamma$ is left free. As a result, according to Eq. (3.25), $V$ is approximated by

$$V_5 = V_0 + \frac{Z_1 Z_2 e^2}{d}\left[\gamma(\xi^2 + \eta^2) + \xi^3 - 3\xi\eta^2\right]. \tag{3.29}$$

This expression is invariant under the symmetry operations relative to the channel center from the $C_{3v}$ point group [124].

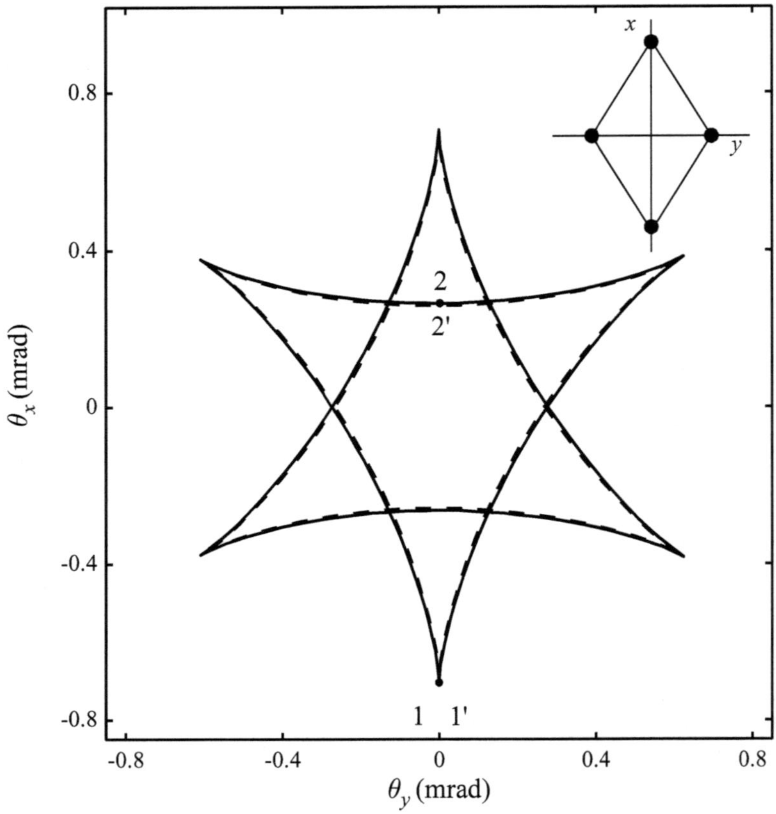


**Fig. 3.24** Rainbow line in the TA plane for 10 MeV protons and a 100-nm thick <111> W crystal obtained by the function $F$ (*solid line*) and the function $G_5$ (*dashed line*) [90]. Points 1 and 2 belong to the former line and points 1′ and 2′ to the latter line. The arrangement of atomic strings defining the channel is shown too

Figure 3.24 gives the rainbow lines in the TA plane in the case of <111> W crystal generated by the functions $F$ and $G_5$, which should be compared with Fig. 3.19. The arrangement of atomic strings defining the channel is shown in the figure. The latter rainbow line is obtained by a fitting procedure in which point 1′ coincides with 1. The resulting distance between points 2′ and 2 is very small. One can see that the rainbow line generated by $V_5$ is a very good approximation of the line obtained by $V$ – much better than the line generated by $V_K$. The application of the same fitting procedure to the remaining 27 cases of <111> crystals has shown that the shape of $\gamma$ as a function of $Z_2$ is very similar to the shape of the density of the <111> atomic strings, $n_\rho$, as a function of $Z_2$. As expected, the conclusion is that the value of the free control variable of the elementary catastrophe is proportional to the value of $n_\rho$ of the chosen very thin crystal.

## 3.3 Zero-Degree Focusing Effect

It has been described in Subsect. 3.1.2 that Krause et al. [44] revealed a scaling law in ion channeling in thin crystals, according to which an angular distribution of transmitted ions changes periodically with the increase of the reduced crystal thickness, $\Lambda$, defined by Eq. (3.16). However, they were concentrated mostly on the first cycle of this change, ending at $\Lambda = 0.5$. Following those results, we decided to explore in detail the evolution of such an angular distribution in a longer range of $\Lambda$, and began with the variation of the yield of transmitted ions in the vicinity of the origin of the TA plane, i.e., the zero-degree yield of transmitted ions, with $L$ [91, 95]. Here, we are going to present the main results of those studies.

The projectiles are $C^{6+}$ ions, and the target is the <100> channel of a Si crystal [91]. The incident ion kinetic energy is $E_0 = 25$ MeV, while the crystal thickness, $L$, is varied from 0 to 5000 atomic layers. The $z$ axis of the reference frame coincides with the channel axis, and the origin lies in the entrance plane of the crystal. The $x$ and $y$ axes of the reference frame are the vertical and horizontal axes, respectively. The incident ion momentum vector is parallel to the channel axis. We adopt Lindhard's ion-atom interaction potential, defined by Eq. (3.10), and the continuum approximation [66], i.e., we use the continuum interaction potential of the ion and $i$th atomic string of the crystal given by Eq. (3.11), with $Z_1 = 6$, $Z_2 = 14$, and $d = 0.543082$ nm [107]. As it has been anticipated above, $L$ will be expressed in atomic layers; in the case under consideration, one atomic layer equals $d$. If expressed in micrometers, the chosen range of $L$ is from 0 to 2.7 μm. Instead of $a_{TF}$, we employ Lindhard's ion-atom screening length, $a_L$, determined by Eq. (2.6). The fitting parameter has been chosen to be $C_L = 4.68$, making the ion-atom interaction potential at the distance coinciding with the distance between an atomic string defining the channel and the point in the IP plane corresponding to the apex of the cusp of the rainbow line in the TA plane (see Fig. 3.12) equal to the corresponding Thomas-Fermi interaction potential [126]. The atomic strings lie on the $x$ and $y$ axes. Their number is $M = 36$, i.e., we include the atomic strings lying on the three square coordination lines nearest to the channel axis. This value gives the average electron density on the channel axis with the accuracy higher than $3 \times 10^{-2}$ [126]. The thermal vibrations of the crystal's atoms are included, via Eq. (3.13), with the one-dimensional atomic thermal vibration amplitude being $\sigma_{th} = 0.00744$ nm [108, 109]. The ion collisions with the crystal's electrons are taken into account partly. We include the ion energy loss but neglect the uncertainty of its transmission angle caused by these collisions. The (electronic) ion energy loss rate can be taken to be

$$-\frac{dE}{dz} = -\frac{Z_1^2}{2}\left(\frac{dE}{dz}\right)^{p,r}, \tag{3.30}$$

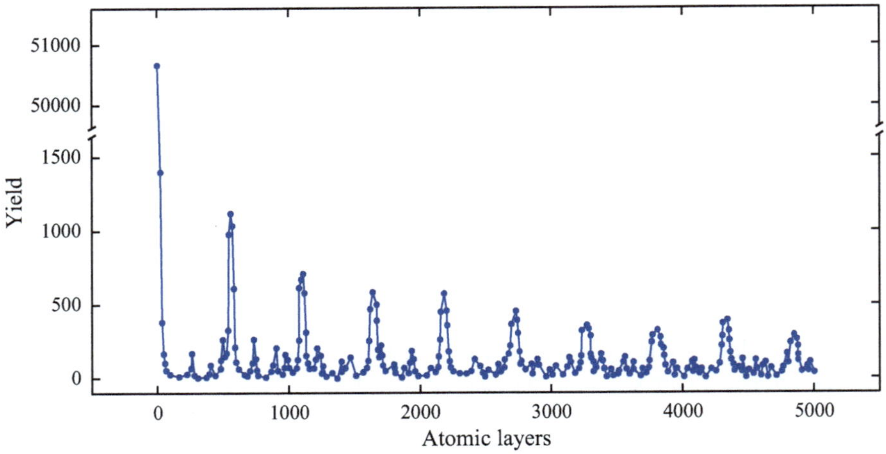


**Fig. 3.25** Dependence of the zero-degree yield of 25 MeV $C^{6+}$ ions transmitted through the <100> channel of a Si crystal on $L$ (expressed in atomic layers) in the case of Lindhard's ion-crystal continuum interaction potential [91]

where $-(dE/dz)^{p,r}$ is the (electronic) energy loss rate of protons in a random Si crystal [44]. For this variable, as in Ref. [44], we use the expression given by Anderson and Ziegler [127].

The angular distributions of transmitted ions are generated as it has been explained in Subsect. 2.3.2. The ion equations of motion in the transverse plane are solved using the Runge-Kutta method of the fourth order [128]. The transverse components of the initial ion position vector, i.e., the components of its impact parameter vector, are chosen uniformly from the uniform distribution within the region of the channel. The number of incident ions is 50,625. Figure 3.25 shows the dependence of the zero-degree yield of transmitted ions on $L$ (expressed in atomic layers). For the region in the TA plane in the vicinity of the origin, we take the region in which the $x$ and $y$ components of the ion transmission angle, $\Theta_x$ and $\Theta_y$, respectively,[2] are smaller than 0.0005°. One can see clearly that the zero-degree yield changes periodically with the increase of $L$. Its nine maxima correspond to the values of $\Lambda$ equal to 0.5, 1.0, ..., and 4.5 [44], i.e., the ends of its first, second, ..., and ninth cycles. The average distance between the neighboring maxima of this dependence is 538 atomic layers, with the standard deviation of 6 atomic layers. The resulting frequency of ion oscillations around the channel axis is $f_L(C^{6+}) = 3.43 \times 10^{13}$ Hz. This effect is called the effect of zero-degree focusing of channeled ions. It is evident that the effect decays with $L$. Analysis has shown that this decay can be reproduced by a sum of two exponential functions. The resulting characteristic half-thicknesses of the crystal are 106 and 1824 atomic

[2] In Sects. 3.3 and 3.4, and Chaps. 4 and 5, unlike in Sects. 3.1 and 3.2, and Chap. 6 , the vertical and horizontal components of the ion transmission angle are denoted by $\Theta_x$ and $\Theta_y$, respectively.

layers. The corresponding values of $\Lambda$ are 0.20 and 1.70, respectively. This means that the first of these values belongs to the first cycle of the zero-degree focusing effect and the second value to its fourth cycle. It should be mentioned that the effect has not yet been observed experimentally.

But what are the origins of the periodicity and decay of the zero-degree focusing effect? The answers to these questions were obtained in the study of transmission of $Ne^{10+}$ ions through the <100> channel of a Si crystal in which $L$ was 1 μm and $E_0$ was varied from 1 to 37 MeV per nucleon [95]. The investigation was performed in the same way as the above-described one, but neglecting the (electronic) ion energy loss, which was found to be small. The projectile atomic number and the obtained value of the fitting parameter in Eq. (3.10) were $Z_1 = 10$ and $C_L = 4.76$, respectively. However, in addition to Lindhard's ion-crystal continuum interaction potential, we employed the corresponding catastrophic continuum interaction potential, given by the standard form of Eq. (3.27) with $\gamma = \delta$,

$$V^+_{K,\gamma} = V_0 + \frac{Z_1 Z_2 e^2}{d}[\gamma(\xi^2 + \eta^2) + \xi^4 + K\xi^2\eta^2 + \eta^4]. \tag{3.31}$$

The modulus of the catastrophe, being the $^4X_9$ elementary catastrophe, was $K = -3.75$, the free control variable $\gamma = 0.297$, and the spatial scaling factor $\rho_0 = 2.90$ a.u. The angular distributions of transmitted ions were generated with 20,449 incident ions.

Figure 3.26 depicts the dependence of the zero-degree yield of transmitted ions on $E_0{}^{-1/2}$, being proportional to $v_0^{-1}$, i.e., to the reciprocal magnitude of the incident ion velocity vector, generated with Lindhard's ion-crystal continuum interaction potential. For the region in the TA plane in the vicinity of the origin, we took the

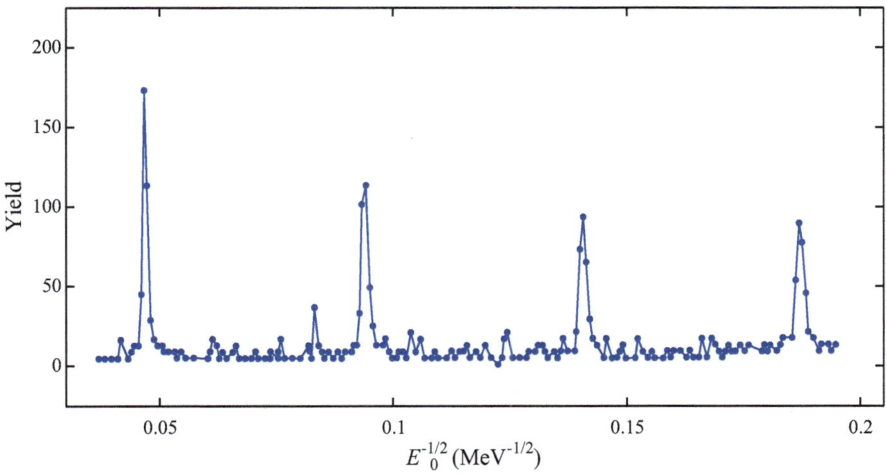


**Fig. 3.26** Dependence of the zero-degree yield of $Ne^{10+}$ ions transmitted through the <100> channel of a 1-μm thick Si crystal on $E_0^{-1/2}$ in the case of Lindhard's ion-crystal continuum interaction potential [95]

region where the $x$ and $y$ components of the ion transmission angle were smaller than 0.0001° × $(60/E_0)^{1/2}$, with $E_0$ given in MeV. The four maxima appearing in this dependence correspond to the values of Λ equal to 0.5, 1.0, 1.5, and 2.0, i.e., the ends of the first, second, third, and fourth cycles of the zero-degree focusing effect. The resulting frequency of ion oscillations around the channel axis is $f_L(Ne^{10+}) = 3.30 \times 10^{13}$ Hz. When the calculations are done with the corresponding catastrophic continuum interaction potential, this variable is $f_K(Ne^{10+}) = 3.39 \times 10^{13}$ Hz. The fact that $f_K(Ne^{10+})$ is very close to $f_L(Ne^{10+})$ is a demonstration of accurateness of the applied catastrophic modeling.

The study was continued with $V^{+}_{K,\gamma}$. The first aim was to see whether the maxima of the zero-degree yield could be connected to a specific region of the IP plane. Therefore, the IP plane was divided in three regions: region I was the region around the channel center, region II the region around the rainbow line in the IP plane, and region III the region around the atomic strings defining the channel. This is shown in Fig. 3.27. The lengths of the sides of the squares separating regions I and II, and

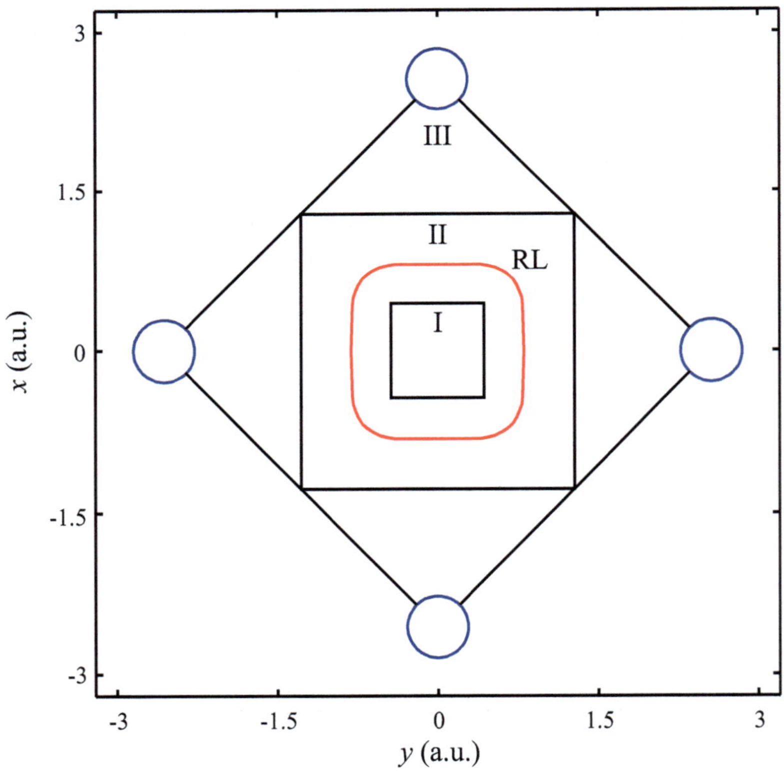


**Fig. 3.27** Arrangement of atomic strings defining the channel in the case of transmission of $Ne^{10+}$ ions through the <100> channel of a Si crystal (*blue color*) [95]. RL – the rainbow line in the IP plane (*red color*). I, II, and III – regions I, II, and III, respectively (*black color*)

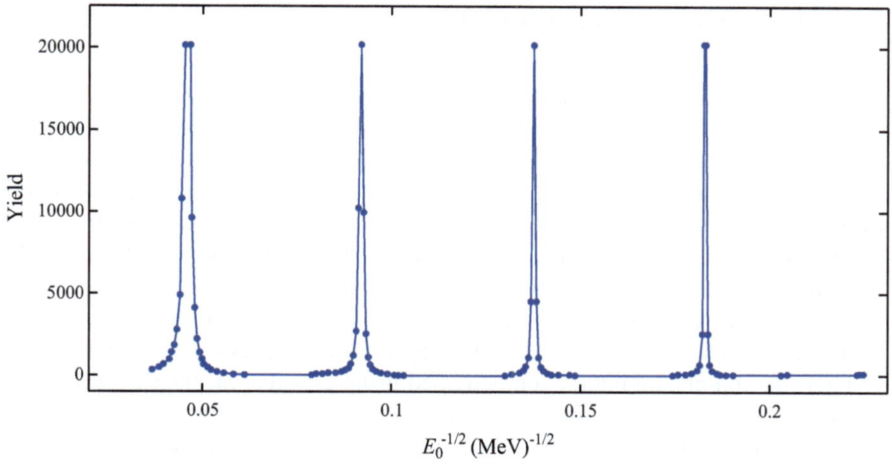


**Fig. 3.28** Zero-degree yield of $Ne^{10+}$ ions transmitted through the <100> channel of a 1-μm thick Si crystal on $E_0^{-1/2}$ when only the harmonic component of the catastrophic ion-crystal continuum interaction potential is taken into account [95]

regions II and III were equal to $a_1/6$ and $a_1/2$, respectively, where $a_1$ was the distance between the atomic strings defining the channel lying on the $x$ or $y$ axis [107]. Analysis clearly demonstrated that the zero-degree focusing effect was to be attributed completely to the ions with the components of the impact parameter vector being in region I, i.e., close to the channel center. Our second aim was to check the role of the harmonic component of $V^{+}_{K,\gamma}$, i.e., its zeroth and second order terms, in making the effect periodic. Figure 3.28 gives the dependence of the zero-degree yield of transmitted ions on $E_0^{-1/2}$ when only the harmonic component of $V^{+}_{K,\gamma}$ is included. The resulting frequency of ion oscillations around the channel axis coincides with $f_K(Ne^{10+})$. Hence, as expected, the conclusion was that the periodicity of the zero-degree focusing effect was to be attributed to the harmonic component of $V^{+}_{K,\gamma}$. The figure shows that in this case, the zero-degree yield does not decay. Consequently, one concludes that the decay of the effect is to be attributed mainly to the anharmonic component of $V^{+}_{K,\gamma}$, i.e., its fourth order terms. If we did not exclude the ion collisions with the crystal's electrons, the decay would be faster.

We also investigated in detail the evolution of the whole angular distribution of 25 MeV $C^{6+}$ ions transmitted through the <100> channel of a Si crystal for $\Lambda < 0.5$[92] as well as of the whole distribution of 60 MeV $Ne^{10+}$ ions transmitted through the same channel for $\Lambda$ between 0.50 and 6.00 [94, 99, 80]. The conclusion was that the cycles of the angular distribution coincided with the cycles of the zero-degree focusing effect. It was established that those cycles were in fact the cycles of change of the crystal rainbow effect. Therefore, they were named the rainbow cycles.

## 3.4 Theory of Crystal Rainbows

We are now ready to present the theory of crystal rainbows [97, 98, 129, 130], which was formulated as a generalization of the model of crystal rainbows [40, 41], described in Subsect. 3.1.1. The generalization was necessary in order to describe the propagation of an ion beam along an axial channel of a crystal that was not necessarily very thin. The theory has been proven to be the proper theory of ion channeling in crystals and nanotubes. This will be explained in detail in the remaining part of this chapter and in Chaps. 4 and 5.

The system we study is an ion moving along an axial crystal channel. As it has been anticipated in Subsect. 2.3.1 and applied in the previous sections of this chapter, the ion-crystal interaction is treated using classical mechanics. The $z$ axis of the reference frame coincides with the channel axis, and the origin lies in the entrance plane of the crystal. The $x$ and $y$ axes of the reference frame are the vertical and horizontal axes, respectively. Often, the incident ion velocity vector is parallel to the channel axis. However, the crystal can be tilted about the $x$ or $y$ axis by an angle smaller than the critical angle for axial channeling, $\psi_c$, defined by Eq. (2.12). In such a case, the incident ion velocity vector makes an angle with the channel axis, being the incident ion angle, which equals the crystal tilt angle. Let us introduce the mappings

$$X = X(x_0, y_0; \Lambda, \varphi) \quad \text{and} \quad Y = Y(x_0, y_0; \Lambda, \varphi), \tag{3.32}$$

and

$$\Theta_x = \Theta_x(x_0, y_0; \Lambda, \varphi) \quad \text{and} \quad \Theta_y = \Theta_y(x_0, y_0; \Lambda, \varphi), \tag{3.33}$$

where $x_0$ and $y_0$ are the transverse components of the initial ion position vector, i.e., the components of its impact parameter vector; $X$ and $Y$ are the transverse components of the final ion position vector; $\Theta_x$ and $\Theta_y$ are the components of the final ion channeling angle, i.e., the components of its transmission angle; $\Lambda$ is the reduced crystal thickness, defined by Eq. (3.16); and $\varphi$ is the incident ion angle. The former mapping is the mapping of the IP plane to the final TP plane, while the latter mapping is the mapping of the IP plane to the TA plane. In order to obtain $X$, $Y$, $\Theta_x$, and $\Theta_y$, the ion equations of motions are solved. One applies either the binary collision model or the continuum string model presented in Subsects. 2.3.3 and 2.3.2, respectively. In the former case, the four variables are obtained upon a three-dimensional following of the ion trajectory, while in the latter case, the calculation is two-dimensional, i.e., the trajectory is followed in the TP plane. Since the ion channeling angle is small, i.e., it is always smaller than $\psi_c$, one can take that $\Theta_x = V_x/V$ and $\Theta_y = V_y/V$, where $V_x = V_x(x_0, y_0; \Lambda, \varphi)$ and $V_y = V_y(x_0, y_0; \Lambda, \varphi)$ are the transverse components of the final ion velocity vector, and $V = V(x_0, y_0; \Lambda, \varphi)$ is its magnitude. The interaction of the ion and a crystal's atom is described by the appropriate interaction potential (see Sect. 2.2).

The theory of crystal rainbows is applied in two steps. In its first step, the spatial and angular distributions of transmitted ions, in the final TP and TA planes, are generated by the computer simulation method described in Sect. 2.3.2.

The second step in applying the theory of crystal rainbows is as follows. Since the components of the ion channeling angle remain small during the whole transmission process, the ion differential transmission cross-section can be taken to be

$$\sigma(x_0, y_0; \Lambda, \varphi) = \frac{1}{|J_\Theta(x_0, y_0; \Lambda, \varphi)|}, \tag{3.34}$$

where

$$J_\Theta(x_0, y_0; \Lambda, \varphi) = \partial_{x_0}\Theta_x\partial_{y_0}\Theta_y - \partial_{y_0}\Theta_x\partial_{x_0}\Theta_y \tag{3.35}$$

is the Jacobian of the functions $\Theta_x(x_0, y_0; \Lambda, \varphi)$ and $\Theta_y(x_0, y_0; \Lambda, \varphi)$, $\partial_{x_0} \equiv \partial/\partial_{x_0}$, and $\partial_{y_0} \equiv \partial/\partial_{y_0}$. The variable $J_\Theta$ is the ratio of the infinitesimal surfaces in the TA and IP planes. It determines the mapping described by Eq. (3.33). Hence, the equation

$$J_\Theta(x_0, y_0; \Lambda, \varphi) = 0 \tag{3.36}$$

gives the angular rainbow lines in the IP plane, i.e., the lines in this plane along which $\sigma$ is infinite. The images of these lines determined by $\Theta_x(x_0, y_0; \Lambda, \varphi)$ and $\Theta_y(x_0, y_0; \Lambda, \varphi)$ are the rainbow lines in the TA plane. On the other hand, the mapping described by Eq. (3.32) is determined by the Jacobian of the functions $X(x_0, y_0; \Lambda, \varphi)$ and $Y(x_0, y_0; \Lambda, \varphi)$. This variable reads

$$J_P(x_0, y_0; \Lambda, \varphi) = \partial_{x_0}X\partial_{y_0}Y - \partial_{y_0}X\partial_{x_0}Y. \tag{3.37}$$

It is the ratio of the infinitesimal surfaces in the final TP and IP planes. Thus, the equation

$$J_P(x_0, y_0; \Lambda, \varphi) = 0 \tag{3.38}$$

gives the spatial rainbow lines in the IP plane. The images of these lines determined by $X(x_0, y_0; \Lambda, \varphi)$ and $Y(x_0, y_0; \Lambda, \varphi)$ are the rainbow lines in the final TP plane. The rainbow lines in the final TP and TA planes separate the bright and dark regions in these planes, which are the bright and dark sides of the spatial and angular crystal rainbows, respectively. It is important to realize that these lines are the envelopes of the ion beam in these planes [131], analogous to the envelope of the outgoing light rays from class III mentioned in Sect. 2.1. As in the first step, one can include in these calculations the thermal vibrations of the crystal's atoms. However, in order to see the rainbow patterns as clear as possible, i.e., not smeared by the ion collisions with the crystal's electrons, the calculations are performed without taking into account these collisions. It should be noted that, usually, one explores either an angular distribution of transmitted ions and the associated rainbow pattern in the TA plane, i.e., the angular crystal rainbow effect, or a spatial distribution of transmitted ions and the associated rainbow pattern in the final TP plane, i.e., the spatial crystal rainbow effect.

Now, we are going to present two applications of the theory of crystal rainbows. In both cases, the projectiles are 60 MeV $Ne^{10+}$ ions, the target is a Si crystal, and the angular crystal rainbow effect is investigated. In the former case, the ions are transmitted through the <100> channel, which is a square channel containing one atomic string of the crystal [97]. In the latter case, the ions are transmitted through the <111> channel, which is a hexagonal channel containing one atomic string [98]. This channel comprises two equilateral triangular subchannels, which are treated here as two separate channels. The ion-atom interaction is described by Moliére's approximation, given by Eq. (3.15), as it was done by Krause et al. [44]. We apply the continuum string model. According to Eqs. (2.13), (2.14), (2.15), and (2.16), the resulting Moliére's interaction potential of the ion and $i$th atomic string reads

$$V_{Mi} = \frac{2Z_1Z_2e^2}{d}\sum_{\ell=1}^{3}\alpha_\ell K_0\left(\frac{\beta_\ell \rho_i}{a_{TF}}\right), \tag{3.39}$$

with $Z_1 = 10$, $Z_2 = 14$, and $d = 0.543082$ nm [107]. Since it has been confirmed in Sect. 3.3 that the periodicity of an angular distribution of transmitted ions is determined by the harmonic component of the continuum potential in the channel, the frequency of ion oscillations around the channel or subchannel axis, appearing in Eq. (3.16), is calculated from the second-order terms of the Taylor expansion of this potential in the vicinity of the channel or subchannel axis, i.e., $f = f_h$. In case of the <100> channel, the crystal thickness is varied from 105 to 632 atomic layers, with one atomic layer being equal to $d$. This range corresponds to the range of $\Lambda$ from 0.10 to 0.60, i.e., from the beginning of the first rainbow cycle to the beginning of the second rainbow cycle. The atomic strings defining the channel lie on the $x$ and $y$ axes (see Fig. 3.27). Their number is $M = 36$, i.e., we take into account the atomic strings lying on the three nearest square coordination lines. In case of the <111> channel, the crystal thickness is changed between 159 and 478 atomic layers, corresponding to the range of $\Lambda$ between 0.10 and 0.45, i.e., within the first rainbow cycle. The atomic strings defining the channel lie on the $x$ and $y$ axes, and the subchannel axes lie on the $x$ axis (see Fig. 3.19 or 3.24). In order to minimize the computation time, the calculations are performed only for the subchannel whose axis lies on the positive part of the $x$ axis. The contribution of the subchannel whose axis lies on the negative part of the $x$ axis is obtained using the fact that the channel is symmetric relative to the $y$ axis. The number of atomic strings is $M = 36$, i.e., we take into account the atomic strings lying on the three nearest triangular coordination lines. The thermal vibrations of the crystal's atoms are introduced by the expression analogous to Eq. (3.13) [132], with the one-dimensional atomic thermal vibration amplitude being $\sigma_{th} = 0.00744$ nm [108, 109]. The ion energy loss caused by its collisions with the crystal's electrons has been found to be small, and it is neglected. The ion equations of motion in the transverse plane are solved numerically [128]. The components of the ion impact parameter vector are chosen uniformly within the region of the channel or subchannel. In case of the <100> channel, the number of incident ions is 502,681, while in case of the <111> ions, it is 260,240. The Jacobian of $\Theta_x$ and $\Theta_y$ [Eq. (3.35)] as well as the angular rainbow lines in the IP plane [Eq. (3.36)] are determined numerically too [128].

Figure 3.29 shows the angular distributions of 60 MeV $Ne^{10+}$ ions transmitted through the <100> channel of a Si crystal for six characteristic values of Λ: 0.10,

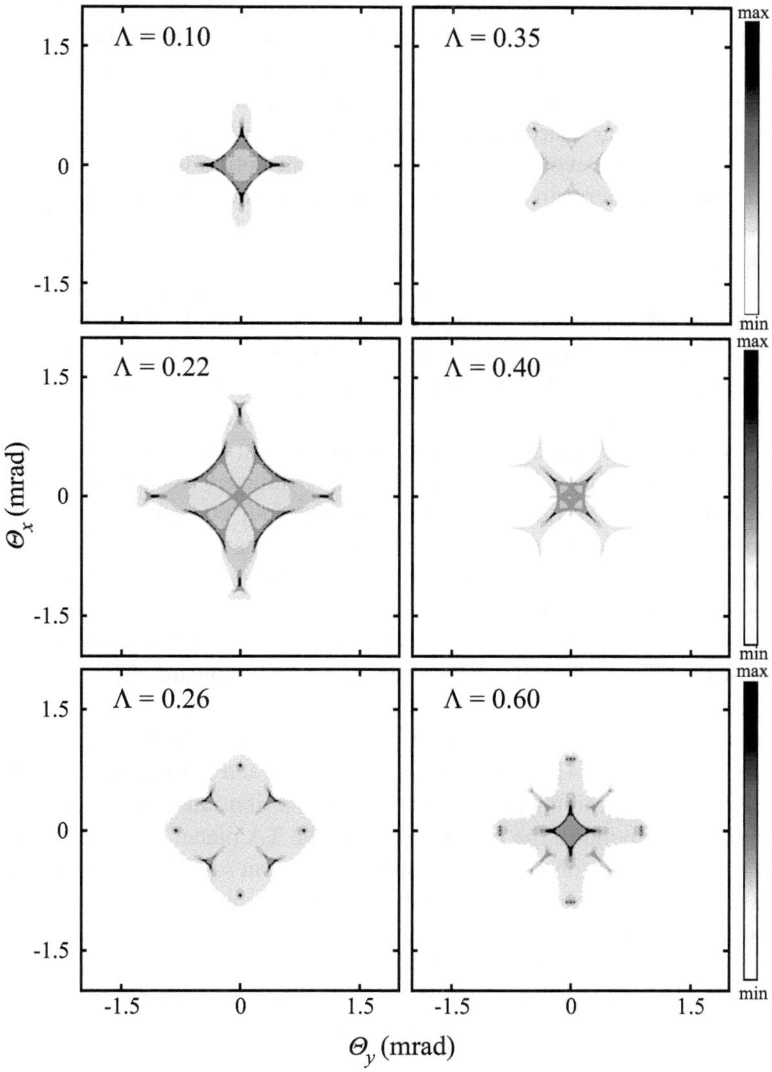


**Fig. 3.29** Angular distributions of 60 MeV $Ne^{10+}$ ions transmitted through the <100> channel of a Si crystal for Λ = 0.10, 0.22, 0.26, 0.35, 0.40, and 0.60. The areas in which the yields of trasmitted ions are above 60, 30, 15, and 5% of the maximal yield are designated by the decreasing tones of *gray color*, respectively (Adapted from Ref. [97])

0.22, 0.26, 0.35, 0.40, and 0.60 [97]. The areas in which the yields of transmitted ions are above 60, 30, 15, and 5% of the maximal yield are designated by the decreasing tones of gray color, respectively. The first five values of $\Lambda$ correspond to the first rainbow cycle while its sixth value corresponds to the beginning of the second rainbow cycle. The angular distribution for $\Lambda = 0.10$ is characterized by four pronounced maxima lying on the lines directed toward the atomic strings defining the channel. For $\Lambda = 0.22$ and 0.26, the angular distributions contain four pronounced maxima lying on the lines directed toward the atomic strings and eight pronounced maxima lying close to the lines directed between the strings. The angular distributions for $\Lambda = 0.35$ and 0.40 are characterized by four pronounced maxima lying on the lines directed between the atomic strings. Finally, for $\Lambda = 0.60$, the angular distribution is similar to the one for $\Lambda = 0.10$, demonstrating the periodicity of the evolution of the angular distribution with an increase of $\Lambda$.

The rainbow patterns in the TA plane corresponding to the angular distributions of transmitted ions shown in Fig. 3.29 are depicted in Fig. 3.30. They contain only the rainbow lines connecting the pronounced maxima of the angular distributions. For $\Lambda = 0.10$, the rainbow pattern consists of a cusped square with the cusps directed toward the atomic strings defining the channel. Since the crystal in question is very thin, this rainbow line would practically coincide with the line generated using the model of crystal rainbows, described in Subsect. 3.1.1 (see Fig. 3.12). When $\Lambda = 0.22$, there are two cusped rectangular rainbow lines lying along the lines directed between the atomic strings, and four cusped isosceles triangular rainbow lines lying along the lines directed toward the strings. For $\Lambda = 0.26$ and 0.35, the rainbow pattern contains four cusped isosceles triangular lines and four points. However, in the former case, the rainbow lines lie along the lines directed between the atomic strings and the rainbow points on the lines directed toward the strings, while in the latter case, the disposition of rainbow lines and points is opposite. The rainbow pattern for $\Lambda = 0.40$ consists of a cusped square with the cusps directed between rather than toward the atomic strings. Finally, when $\Lambda = 0.60$, the rainbow pattern again contains a cusped square with the cusps directed toward the atomic strings. For $\Lambda = 0.50$, the rainbow line reduces to a rainbow point [114]. Comparison of Figs. 3.29 and 3.30 clearly demonstrates that all the pronounced maxima of the angular distributions, except those lying at the origin, correspond to the rainbow lines in the TA plane. This means that their origin is the crystal rainbow effect. Therefore, one can say that the evolution of the angular distribution with an increase of $\Lambda$ is fully determined by the evolution of the angular crystal rainbow pattern.

Figures 3.31(a) and (b) give the yields of 60 MeV $Ne^{10+}$ ions transmitted through the <100> channel of a Si crystal for $\Lambda = 0.22$ along the lines directed toward and between the atomic strings defining the channel, respectively. One can see that the maximum of the yield shown in the former figure denoted by 1 corresponds to the effect of zero-degree focusing of channeled ions, that maxima 2 correspond to the intersections of the two cusped rectangular rainbows with the $\Theta_x$ axis, and that maxima 3 and 4 correspond to the intersections of the two cusped isosceles triangular rainbows with the same axis. One can also see that the maximum of

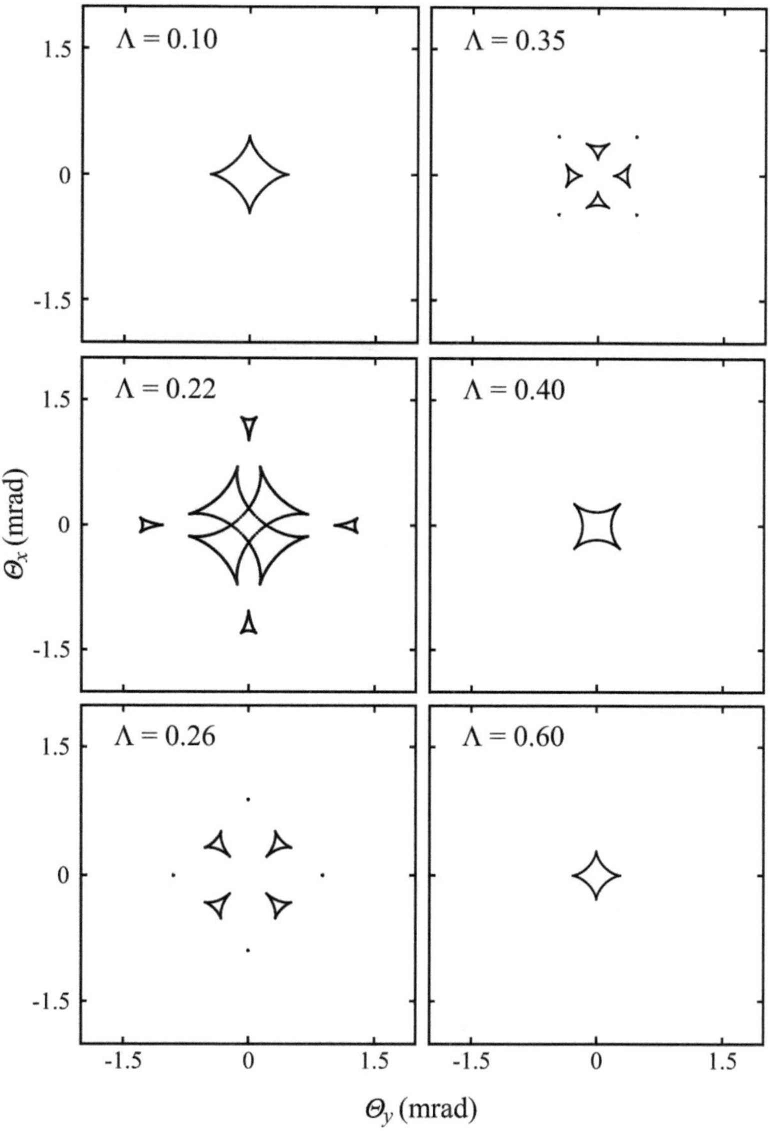


**Fig. 3.30** Rainbow lines in the TA plane in the transmission of 60 MeV $Ne^{10+}$ ions through the <100> channel of a Si crystal for $\Lambda = 0.10$, 0.22, 0.26, 0.35, 0.40, and 0.60 [97]

the yield shown in the latter figure denoted by 1 corresponds to the zero-degree focusing effect, and that the maxima 2 and 3 correspond to the intersections of the two cusped rectangular rainbows with the line $\Theta_y = \Theta_x$. It must be noted that in an experiment, the widths of these maxima would be larger – primarily due to the uncertainty of the ion transmission angle caused by its collisions with the crystal's electrons and a finite resolution of the ion detector. In order to observe these

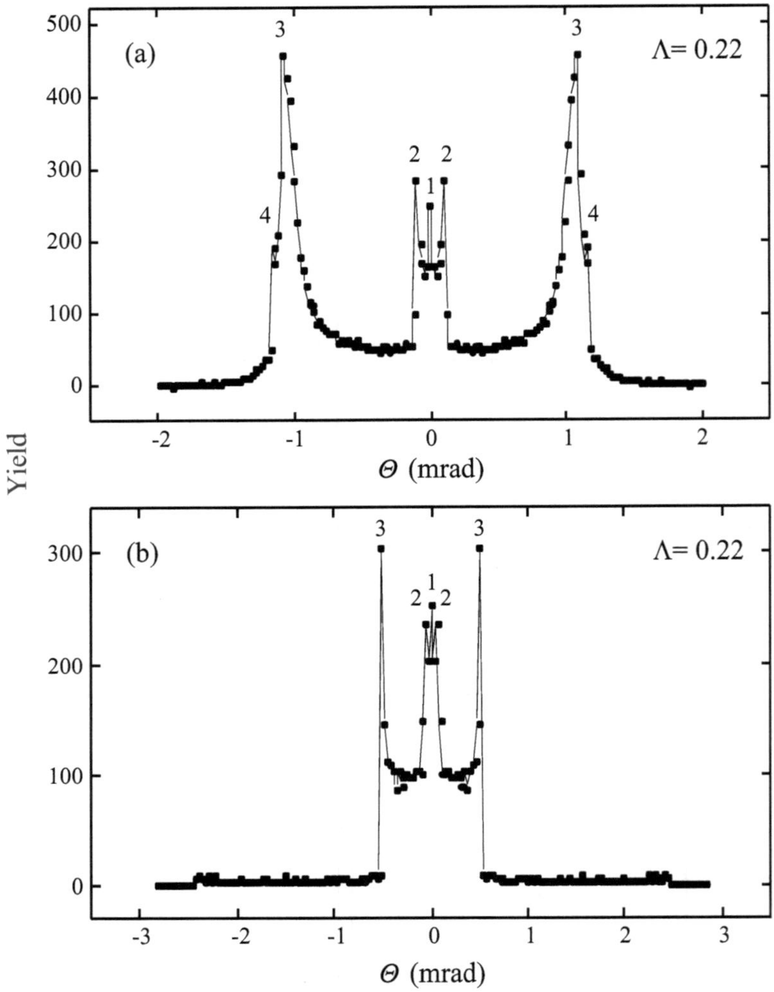


**Fig. 3.31** Yields of 60 MeV $Ne^{10+}$ ions transmitted through the <100> channel of a Si crystal for $\Lambda = 0.22$ (**a**) along the line directed toward the atomic strings defining the channel and (**b**) along the line directed between the atomic strings. $\Theta = \left(\Theta_x^2 + \Theta_y^2\right)^{1/2}$ is the ion transmission angle [97]

maxima, one must perform a measurement similar to the latter measurements that have been described in Subsect. 3.1.2 [44], and the measurements some of which will be presented in Subsect. 4.1.1 [45–48].

Figure 3.32 contains the rainbow patterns in the TA plane for 60 MeV $Ne^{10+}$ ions transmitted through the <111> channel of a Si crystal for six characteristic values of $\Lambda$: 0.15, 0.20, 0.25, 0.30, 0.34, and 0.45 [98]. They contain only the rainbow lines connecting the pronounced maxima of the angular distributions of transmitted ions. For $\Lambda = 0.15$, there are two cusped equilateral triangular rainbow lines with the cusps lying along the lines $\Phi = 2n\pi/3$ and $\Phi = (2n+1)\pi/3$, $n$=0–2.

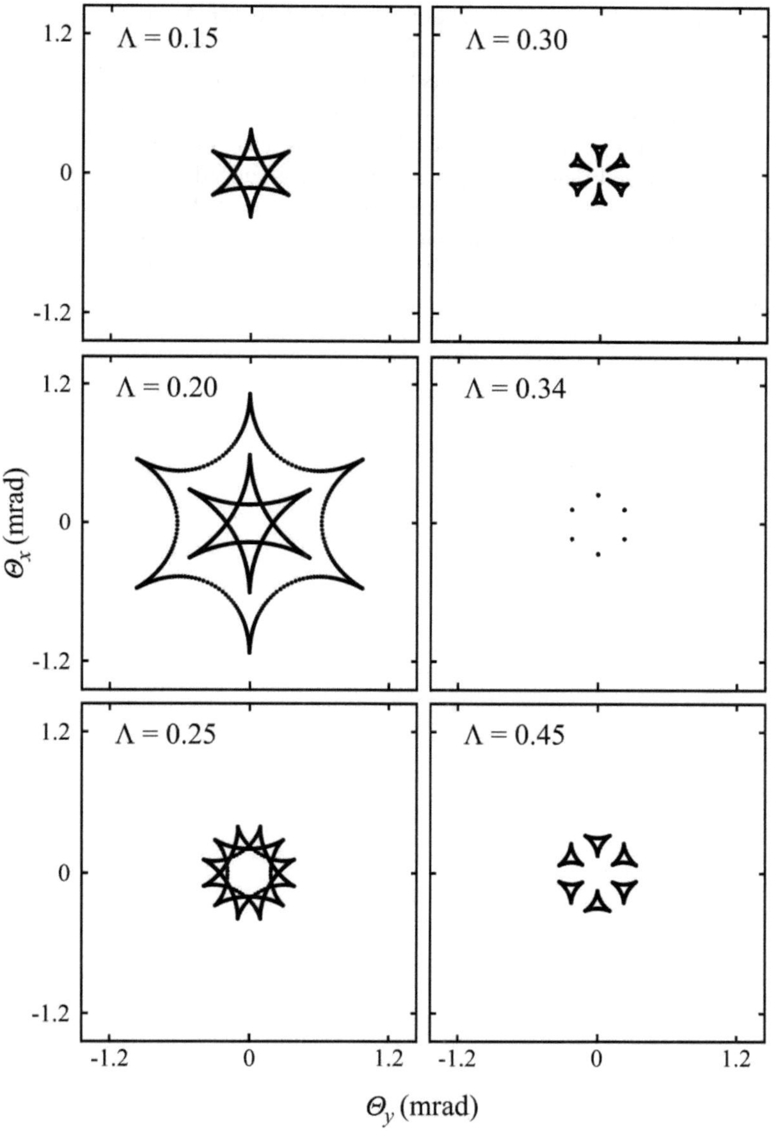


**Fig. 3.32** Rainbow lines in the TA plane for the transmission of 60 MeV $Ne^{10+}$ ions through the <111> channel of a Si crystal for $\Lambda = 0.15$, 0.20, 0.25, 0.30, 0.34, and 0.45 [97]

This rainbow line would practically coincide with the line obtained using the model of crystal rainbows, described in Subsect. 3.2.2 (see Fig. 3.19 or 3.24). The rainbow pattern for $\Lambda = 0.20$ contains two cusped triangular lines corresponding to the ones appearing for $\Lambda = 0.15$ and a cusped equilateral hexagonal line with the cusps lying on the lines $\Phi = n\pi/3$, $n = 0-5$. For $\Lambda = 0.25$, there are three cusped rectangular rainbow lines with the sides perpendicular to the lines $\Phi = 3n\pi/6$, $\Phi = (3n+2)\pi/6$, and $\Phi = (3n + 4)\pi/6$, $n = 0$–3. The rainbow pattern for $\Lambda = 0.30$ consists of six cusped isosceles triangular lines lying along the lines $\Phi = n\pi/3$, $n = 0$–5. For $\Lambda = 0.34$, there are six rainbow points lying on the lines $\Phi = n\pi/3$, $n = 0$–5. Finally, the rainbow pattern for $\Lambda = 0.45$ contains six cusped isosceles triangular lines similar to the ones appearing for $\Lambda = 0.30$. As in case of the <100> channel, it has been established that the origin of all the pronounced maxima of the angular distributions of transmitted ions, except those lying at the origin, is the crystal rainbow effect. Again, one can state that the evolution of the angular crystal rainbow pattern with an increase of $\Lambda$ fully determines the evolution of the angular distribution.

# Chapter 4
# Rainbows in Proton Channeling in Silicon Crystals

In this chapter, some of the high-resolution proton channeling measurements with a very thin (100) Si crystal conducted by the Singapore group [45–48] will be thoroughly analyzed. Those extraordinary measurements have proven to be crucial for verification of the theory of crystal rainbows as the proper theory of ion channeling in crystals. We shall present the results of three experimental and theoretical studies that were induced by those measurements and performed jointly by us and the Singapore group. The first study leads to the very accurate ion-atom interaction potentials [133]. As a continuation, we shall additionally explore the process that is inverse to the transmission process under consideration, and, thus, fully answer the question of its multiplicity, which is directly connected to the essence of the crystal rainbow effect, being the ion focusing along a line reflecting the symmetry of the illuminated crystal. The second study is devoted to the effect of superfocusing of channeled ions, which is the effect of spatial focusing occurring in the middle of each rainbow cycle [134]. The third study contains the proof that the doughnut effect in ion channeling, occurring with tilted crystals, which has been seen and measured many times, is in fact a crystal rainbow effect [130]. Before presenting the results of the second and third studies, we shall describe in detail the superfocusing and doughnut effects, respectively.

In each of the three studies, the $z$ axis of the reference frame coincides with the channel axis, and the origin lies in the entrance plane of the crystal. The $x$ and $y$ axes of the reference frame are the vertical and horizontal axes, respectively.



N. Nešković et al., *Rainbows in Channeling of Charged Particles in Crystals and Nanotubes*, Lecture Notes in Nanoscale Science and Technology 25,
DOI 10.1007/978-3-319-61524-0_4

## 4.1 Rainbow Interaction Potentials

### *4.1.1 High-Resolution Measurements of the Crystal Rainbows*

In ion channeling experiments in which angular distributions of ions transmitted through a crystal are looked for, the measurement resolution is determined by: (i) the mass and charge, the kinetic energy and its spread, and the divergence or convergence angle of the incident ion beam, (ii) the diameter of the beam spot at the crystal surface, (iii) the crystal thickness, alignment, and quality, (iv) the thermal vibrations of the crystal's atoms and the ion collisions with the crystal's electrons, and (v) the detector characteristics and its distance from the crystal.

In the period between 2011 and 2014, the Singapore group reported on a series of high-resolution transmission measurements with protons of incident kinetic energy, $E_0$, of 2.0 MeV focused to a spot with a diameter of about 1 μm on the surface of a 55-nm thick (100) Si crystal [45–48]. The experiments were performed with a goniometer that made the protons propagate along the <100>, <110>, and <111> axial crystal channels, being the major axial channels, and along several minor axial channels. The angular distributions of transmitted protons were also measured with the crystal tilted away from those crystallographic directions. For the protons moving along the <100> channels, the reduced crystal thickness was $\Lambda = 0.12$, meaning that, in that case, the crystal was very thin. The proton beam current was about 10 pA, and its convergence angle about 0.01°. The crystal was fabricated using the buried $SiO_2$ layer in a silicon-on-insulator wafer as a stop for KOH etching through a backside opening in a protective polymer layer, producing a crystal surface area of about 0.5 $mm^2$ with a roughness of about 0.4 nm. The distributions were recorded by photographing a highly sensitive aluminum-coated YAG scintillator screen. Its distance from the crystal was 50 cm. The camera exposure time was 0.8 s. The obtained results were successfully reproduced using the FLUX computer simulation code [83], mentioned in Subsect. 2.3.3, and the ZBL proton-atom interaction potential [74–76], described in Sect. 2.2.

The advantages of those measurements relative to the two measurements described in Subsect. 3.1.2 [42–44] from the point of view of resolution are connected to the facts that the crystal was considerably thinner (55 nm versus 140 and 179 nm) and the proton beam spot diameter much smaller (1 μm versus 0.5 and 0.18 mm). In the former previous experiment, the values of Λ were close to 0.25, and in the latter experiment, they were between 0.25 and 1. On the other hand, the measurements in question were qualitative, unlike the previous measurements, which had been quantitative.

### 4.1.2 Interaction Potentials

In the study to be presented in the next subsection, Molière's and the ZBL ion-atom interaction potentials will be employed. In the case of Molière's interaction potential, we shall begin the consideration with the screening function including Firsov's ion-atom screening length, defined by Eqs. (2.2) and (2.5). We denote this interaction potential as the M($a_F$) potential. The screening function of the ZBL interaction potential is determined by Eqs. (2.7) and (2.8).

The parameters of the M($a_F$) ion-atom interaction potential as well as of the ZBL potential were determined to make them accurate dominantly for small ion-atom distances. As it has been said in Subsect. 3.1.2, Krause et al. [44], in analyzing their experimental results, concluded that they were better reproduced by Molière's interaction potential with the Thomas-Fermi atomic screening length, $a_{TF}$, given by Eq. (2.4), than by the M($a_F$) potential. We denote this interaction potential as the M($a_{TF}$) potential. That conclusion was attributed to the fact that each recorded angular distribution of transmitted ions was generated by the ions moving close to the channel axis, i.e., far from the atomic strings of the crystal defining the channel. One cannot expect an interaction potential that has proven to be accurate close to the atoms of the strings, i.e., for small ion-atom distances, to be accurate close to the channel axis, i.e., for large ion-atom distances, where many atoms influence the ion propagation.

The M($a_{TF}$) ion-atom interaction potential can be written in a form depending on $a_F$, rather than on $a_{TF}$, with parameters ($\beta_\ell$) changed to $\left(\beta_\ell^c\right) = (\beta_\ell a_F / a_{TF})$. For $Z_1 = 1$ and $Z_2 = 14$, $\left(\beta_\ell^c\right) = (5.124, 1.025, 0.2562)$. We denote this interaction potential as the M$^c$($a_F$) potential.

### 4.1.3 Morphological Method of Extraction of the Interaction Potentials

Let us now concentrate on a sequence of high-resolution measurements with 2.0, 1.5, 1.0, and 0.7 MeV focused proton microbeams channeled in a 55-nm thick <100> Si crystal performed by the Singapore groups and analyzed by us [133]. Figure 4.1 shows the measured angular distributions of transmitted protons for $E_0$=2.0 and 0.7 MeV together with the corresponding distributions generated using the first part of the theory of crystal rainbows, described in Sect. 3.4, with two proton-atom interaction potentials, the ZBL and M($a_{TF}$) potentials, and the continuum approximation [66]. The corresponding values of $\Lambda$ calculated for the ZBL interaction potential are 0.12 and 0.21, respectively. The atomic strings of the crystal defining the channel intersect the lines $y = \pm x$. The number of atomic strings included in the calculations is 36, i.e., we take into account the strings lying on the three nearest square coordination lines. In the calculations, we have taken into account the thermal vibrations of the crystal's atoms but not the proton collisions with the

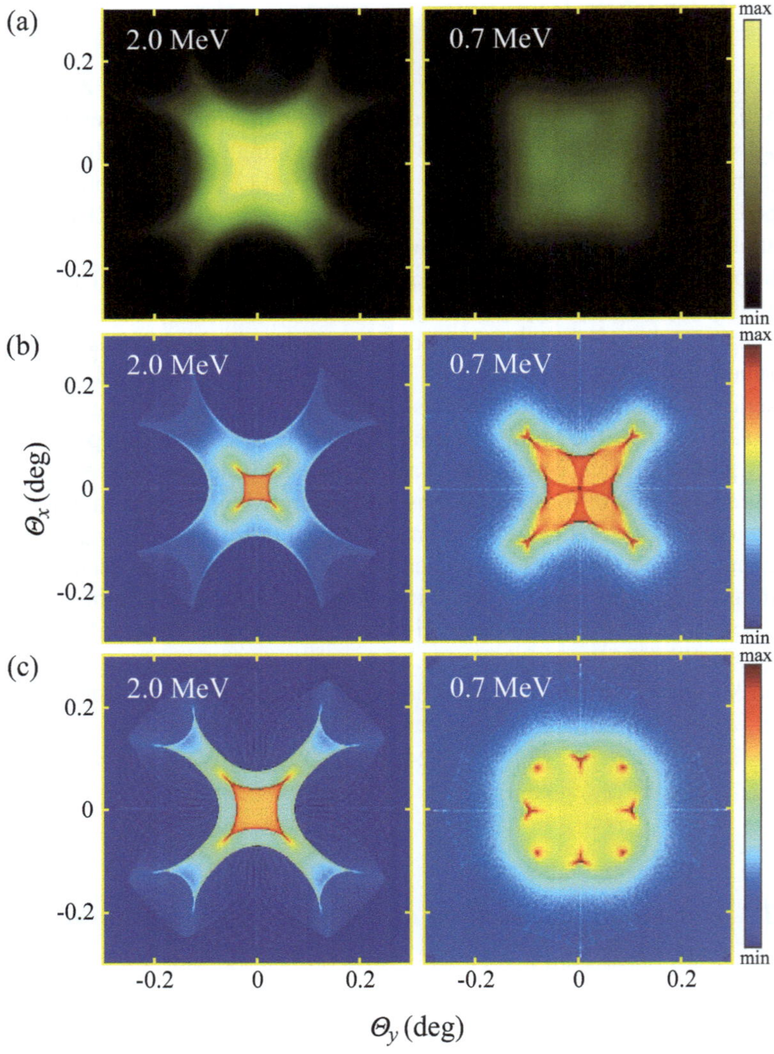


**Fig. 4.1** (**a**) Experimental angular distributions of 2.0 and 0.7 MeV protons transmitted through a 55-nm thick <100> Si crystal [133]. (**b**) Corresponding theoretical angular distributions obtained with the ZBL ion-atom interaction potential [133]. (**c**) Corresponding theoretical angular distributions generated with the M($a_{TF}$) interaction potential [133]

crystal's electrons. It has been found that the contribution of the second effect to the shapes of the distributions is minor. As has been said in Subsect. 4.1.1, the experimental distributions under consideration, as well as the remaining ones given in Figs. 4.4b, e were recorded by photographing a scintillator screen. The technique did not make possible the extraction of sufficiently accurate yields of transmitted protons along different lines on the screen, in order to compare them precisely with the corresponding theoretical yields. The only possibility has been to compare the shapes and extents of the corresponding parts of the experimental and theoretical distributions, i.e., to perform a morphological comparison of the distributions. The central part of each theoretical distribution obtained for $E_0 = 2.0$ MeV follows the shape of a cusped square with four maxima lying on the apices of the cusps. However, the extent of this part of the distribution obtained with the M($a_{TF}$) interaction potential is larger than the extent of the part of the distribution generated with the ZBL potential. The central part of the corresponding experimental distribution also follows the shape of a cusped square. Its extent is closer to the extent of the central part of the distribution generated with the M($a_{TF}$) interaction potential. The peripheral part of each theoretical distribution obtained for this value of $E_0$ follows the shape of a line with four pairs of cusps. The extent of the peripheral part of the corresponding measured distribution is closer to the extent of the peripheral part of the distribution obtained with the ZBL interaction potential.

The theoretical angular distribution depicted in Fig. 4.1 obtained for $E_0 = 0.7$ MeV with the ZBL proton-atom interaction potential also has the shape of a cusped square with four maxima lying on the apices of the cusps. In addition, it has an internal structure composed of four cusped isosceles triangles joined at the origin with four pairs of maxima lying on the apices of the cusps away from the origin and a maximum at the origin. The corresponding theoretical distribution generated with the M($a_{TF}$) interaction potential has approximately the shape of a square with four stronger maxima lying on the lines $\Theta_x = 0$ and $\Theta_y = 0$ and four weaker ones lying on the lines $\Theta_y = \pm\Theta_x$, but without a maximum at the origin. The corresponding experimental distribution has approximately the shape of a square too, being closer to the theoretical distribution obtained with the M($a_{TF}$) interaction potential. Also, it does not have a maximum at the origin, like the theoretical distribution obtained with the M($a_{TF}$) interaction potential and unlike the distribution generated with the ZBL potential.

On the basis of these comparisons, we assume that we can modify the M$^c$($a_F$) proton-atom interaction potential and make it accurate close to the channel axis but without compromising its accuracy close to the atomic strings. We shall construct such an interaction potential and employ it for the analysis of the above-mentioned sequence of high-resolution proton channeling measurements. The construction will be based on adjusting the shapes of the rainbow lines in the TA plane by a minimal modification of the M$^c$($a_F$) interaction potential. This intention is based on the well-established morphological fact that an angular distribution of transmitted ions, generated for an incident ion kinetic energy, crystal thickness, and crystal tilt angle, and its evolution with these parameters are fully determined by the associated rainbow pattern in the TA plane and its evolution [97, 98]. The resulting interaction potential will be denoted as the rainbow potential. Figure 4.2 gives the dependences of the ZBL and M($a_{TF}$) interaction potentials on the proton-atom distance. The inset shows the three components of the M$^c$($a_F$) interaction potential, to be modified in order to obtain the rainbow potential.

The construction of the rainbow proton-atom interaction potential is performed for $E_0 = 2.0$ MeV since, in this case, two well-separated rainbow lines appear in the

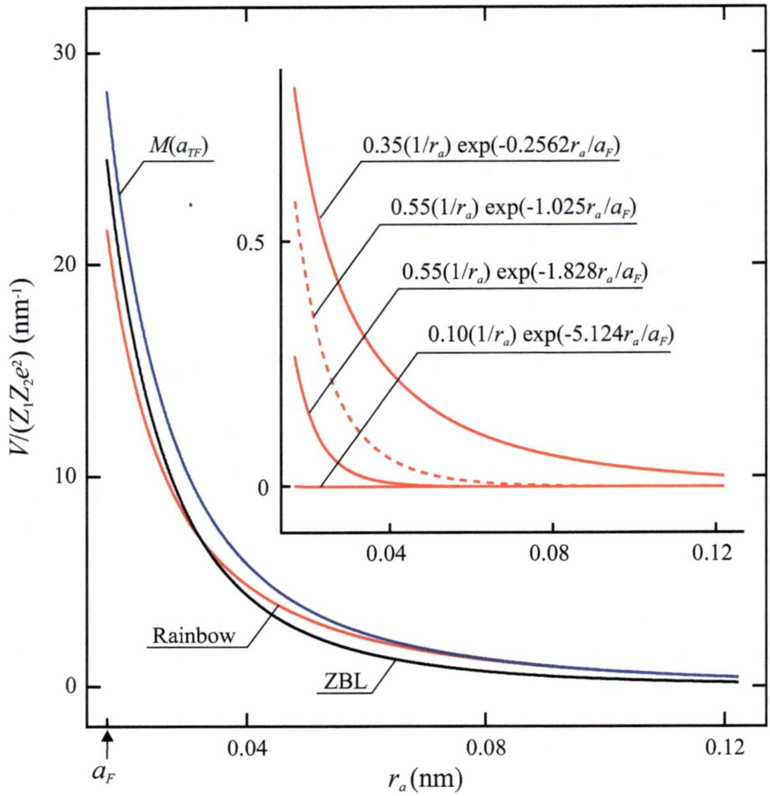


**Fig. 4.2** Dependences of the ZBL, M($a_{TF}$) and rainbow ion-atom interaction potentials on $r_a$ [133]. *Inset*: Components of the M$^c$($a_F$) interaction potential and the second component of the rainbow potential as functions of $r_a$ [133]

IP plane. This is shown in Fig. 4.3a for the ZBL and M($a_{TF}$) interaction potentials. These lines have been obtained using the second part of the theory of crystal rainbows. The thermal vibrations of the crystal's atoms are included. The first line lies close to the channel center and the second one close to the atomic string. It is worth mentioning here that the crystals used by Krause et al. [42–44] were not sufficiently thin for two such (separated) lines to occur in the IP plane. The associated rainbow lines in the TA plane are depicted in Fig. 4.3b. One should also note that, in our earlier study [97], only the central line was registered (see Fig. 3.30). The reason was the fact that the rainbow in question was weak. We now focus on points 1 and 2 in the IP plane, belonging to the first and second rainbow lines, and on points 1′ and 2′ in the TA plane, being the images of points 1 and 2, respectively. Analysis has shown that the positions of points 1′ and 2′ are not sensitive to the changes of the parameter $\beta_1^c$, are sensitive to the changes of the parameter $\beta_2^c$, and are very sensitive to the changes of the parameter $\beta_3^c$. Taking these findings into account, we choose to keep the parameters ($\alpha_\ell$), $\beta_1^c$, and $\beta_3^c$ fixed

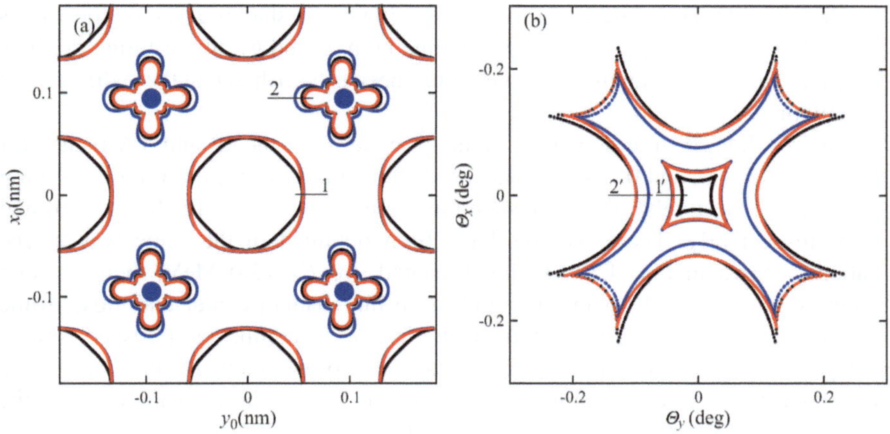


**Fig. 4.3** (**a**) Rainbow patterns in the IP plane for 2.0 MeV protons transmitted through a 55-nm thick <100> Si crystal for the ZBL, M($a_{TF}$), and rainbow ion-atom interaction potentials (*black*, *blue*, and *red lines*, respectively); the *blue line* close to the origin is covered by the *red line* [133]. The *blue full circles* represent the atomic strings defining the channel. (**b**) Associated rainbow patterns in the TA plane; the *blue line* close to the origin is covered by the *red line* [133]

**Table 4.1** Values of the parameters of the rainbow proton-atom interaction potentials in the <100>, <110>, and <111> channels of a Si crystal

| Channel | $\beta_1^r$ | $\beta_2^r$ | $\beta_3^r$ |
|---|---|---|---|
| <100> | 5.124 | 1.828 | 0.2562 |
| <110> | 5.124 | 1.242 | 0.2562 |
| <111> | 5.124 | 1.475 | 0.2562 |

and change the parameter $\beta_2^c$ with the objectives of: (i) moving point 1′ corresponding to the $M^c(a_F)$ interaction potential to minimize its distance from point 1′ corresponding to the M($a_{TF}$) potential, and (ii) moving point 2′ corresponding to the $M^c(a_F)$ potential to minimize its distance from point 2′ corresponding to the ZBL potential. In fact, we minimize the sum of squares of the relative distances between points 1′ and between points 2′. As a result, parameter $\beta_2^c$ is modified from 1.025 to 1.828. The corresponding component of the rainbow interaction potential is given in the inset of Fig. 4.2. Thus, the rainbow interaction potential is defined by Eqs. (2.2) and (2.5) with the parameters ($\beta_\ell$) replaced by $(\beta_\ell^r)$, given in Table 4.1. This interaction potential is presented in Fig. 4.2, and the associated rainbow lines in the IP and TA planes are shown in Figs. 4.3a, b, respectively. The central rainbow line in the TA plane generated with the rainbow interaction potential practically coincides with the central rainbow line generated with the M($a_{TF}$) potential, and the peripheral rainbow line in this plane obtained with the rainbow potential is very close to the peripheral rainbow line obtained with the ZBL potential.

One should mention here that, if the adjusting of the shapes of the rainbow lines in the TA plane were performed by changing both $\beta_3^c$ and $\beta_2^c$, the resulting rainbow potential would be different. However, it has been established that these differences are minor.

Figures 4.4a, d, b, e depict the rainbow patterns in the IP and TA planes for $E_0 = 2.0$, 1.5, 1.0, and 0.7 MeV obtained with the rainbow proton-atom interaction potential, respectively. The corresponding values of $\Lambda$ are 0.16, 0.19, 0.23, and 0.28, respectively. These patterns have been obtained in the same way as the patterns shown in Fig. 4.3. As already stated, for $E_0 = 2.0$ MeV, there are two rainbow lines in the IP plane, lying close to the channel center and close to the atomic string. The rainbow pattern in the TA plane comprises a cusped square, being the image of the line in the IP plane close to the channel center, and a line with four pairs of cusps, being the image of the line in the IP plane close to the atomic string. The rainbow patterns in the IP and TA planes for $E_0 = 1.5$ MeV are similar to those for $E_0 = 2.0$ MeV with the lines in both planes lying closer to each other. The rainbow pattern in the IP plane for $E_0 = 1.0$ MeV contains six lines. There are two equivalent quasi-rectangular lines connecting the neighboring channels, i.e., lying across the channel walls, and four equivalent quasi-elliptical lines lying close to the atomic string. There is no line around the channel center. The rainbow pattern in the TA plane contains two crossed cusped rectangular lines, being the images of the two quasi-rectangular lines in the IP plane, and four cusped isosceles triangular lines, being the images of the four quasi-elliptical lines in the IP plane. Such a pattern was registered in our earlier study [97] (see Fig. 3.30). For $E_0 = 0.7$ MeV, there are two equivalent pairs of quasi-elliptical rainbow lines in the IP plane lying in the neighboring channels close to the channel walls and four equivalent rainbow points in the IP plane lying close to the atomic string. Again, there is no line around the channel center. The rainbow pattern in the TA plane comprises four cusped isosceles triangular lines, being the images of the two pairs of quasi-elliptical lines in the IP plane, and four points, being the images of the four points in the IP plane. Such a pattern was also observed in our earlier study [97] (see Fig. 3.30). In all these cases, the inner side of each rainbow line in the TA plane is the bright side of the rainbow, while its outer side is the dark side of the rainbow. It should be mentioned that for $E_0 = 0.7$ MeV, there is no line in the TA plane around the origin, which explains the above-mentioned fact that the corresponding experimental distribution does not have a maximum at the origin.

Figures 4.4b, e also give the experimental angular distributions of transmitted protons for $E_0 = 2.0$, 1.5, 1.0 and 0.7 MeV. The corresponding rainbow patterns in the TA plane generated with the rainbow proton-atom interaction potential fully determine the distributions – they appear as the "skeletons" of the distributions. A careful inspection and comparison of the measured distributions and rainbow patterns demonstrate excellent agreement. Figures 4.4c, f depict the corresponding angular distributions generated in the same way as the distributions shown in Fig. 4.1, i.e., using the first part of the theory of crystal rainbows, but with the rainbow interaction potential. A comparison of the measured and simulated

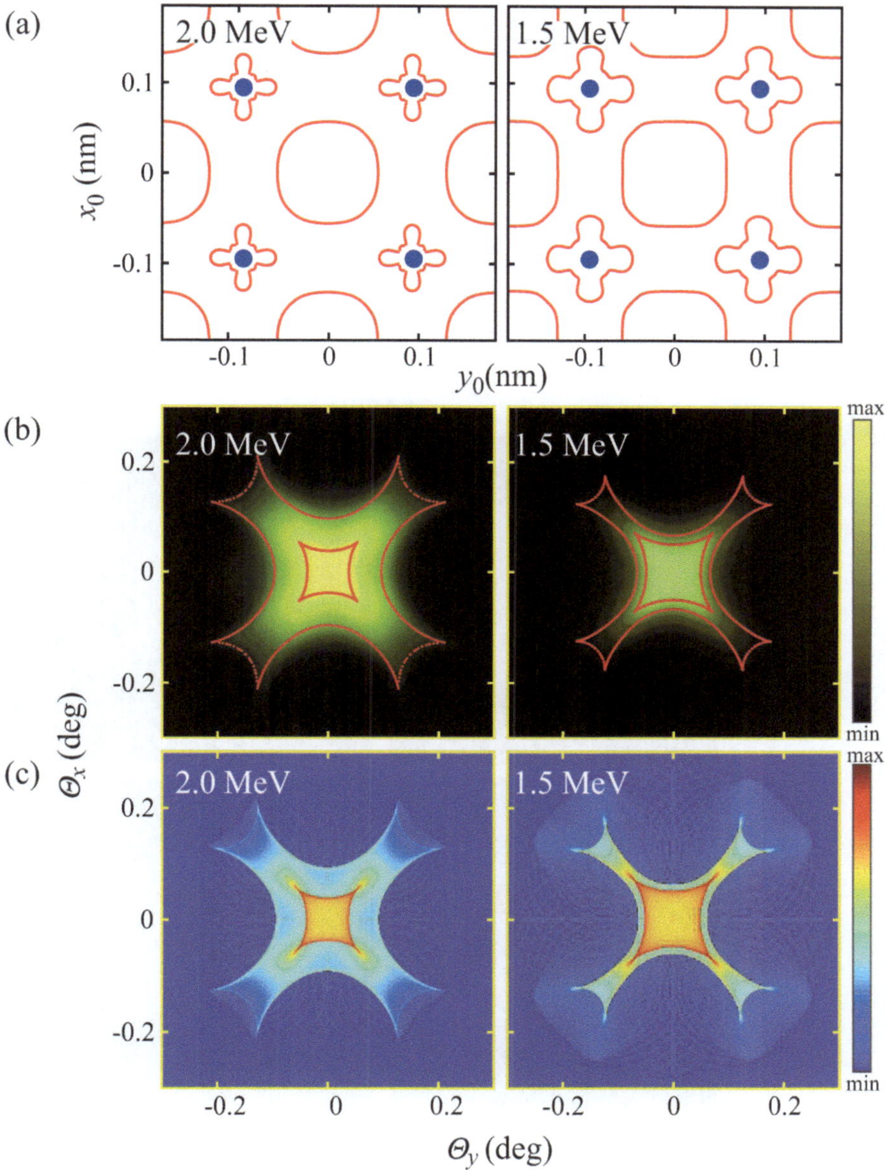


**Fig. 4.4** (**a**) and (**d**) Rainbow patterns in the IP plane for 2.0, 1.5, 1.0, and 0.7 MeV protons transmitted through a 55-nm thick <100> Si crystal generated with the rainbow ion-atom interaction potential. The *blue full circles* represent the atomic strings defining the channel. (**b**) and (**e**) Corresponding experimental angular distributions of transmitted protons and the associated rainbow patterns in the TA plane (*red lines*). (**c**) and (**f**) Associated theoretical distributions generated with the rainbow interaction potential (Adapted from Ref. [133])

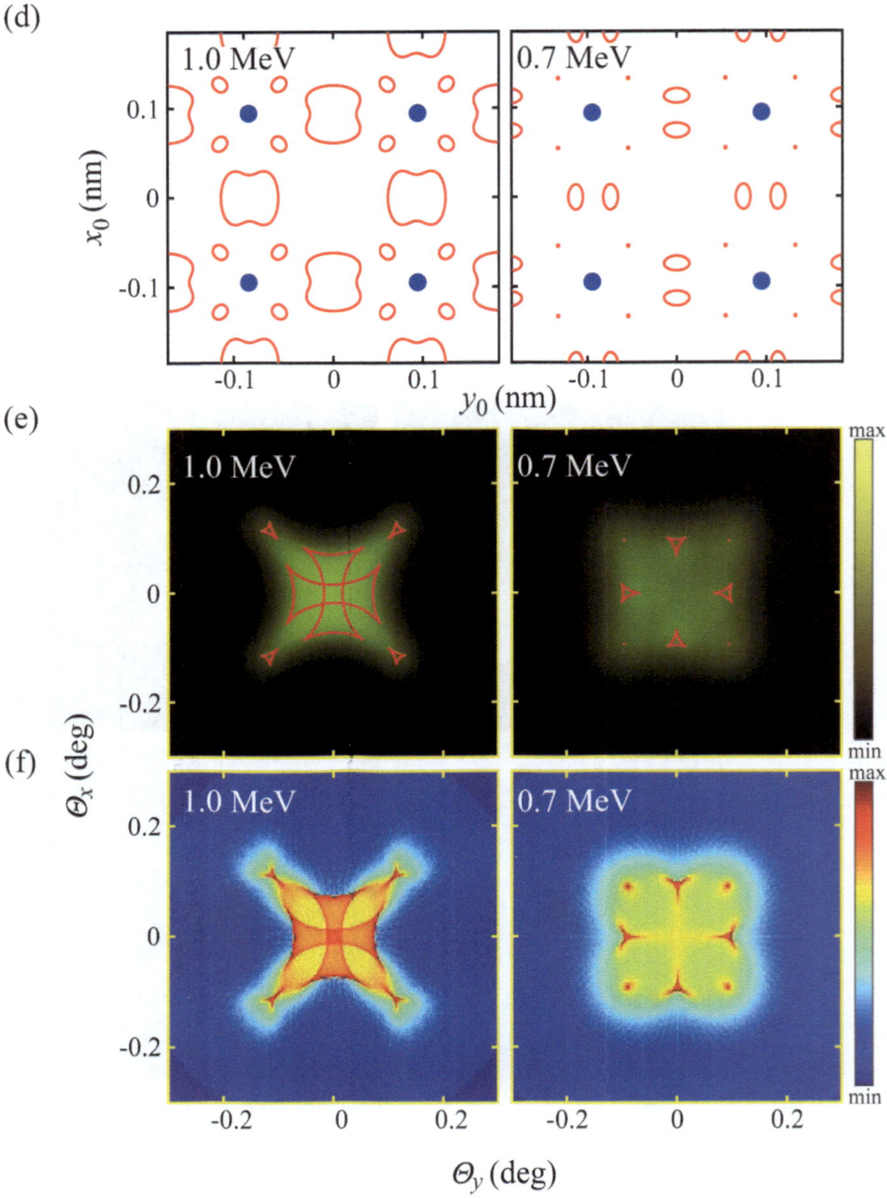


**Fig. 4.4** (continued)

distributions also demonstrates excellent agreement, in spite of the fact that the latter distributions have been obtained using the continuum approximation with the thermal vibrations of the crystal's atoms included but with the focusing of the incident proton beam, the proton collisions with the crystal electrons, and crystal defects neglected. On the basis of these comparisons, we conclude that the above-made assumption on the possibility of modification of the $M^c(a_F)$ interaction potential for protons transmitted through the <100> Si crystal has been proven.

If one starts from the four rainbow patterns given in Figs. 4.4a, d and imagines the evolution of the pattern with a decrease of $E_0$ from 2.0 to 0.7 MeV, it becomes evident that for a certain value of $E_0$ between 1.5 and 1.0 MeV, the rainbow line close to the channel center will join the rainbow lines close to the atomic strings defining the channel. Similarly, for a certain value of $E_0$ between 1.0 and 0.7 MeV, each quasi-rectangular rainbow line connecting the neighboring channels will split into a pair of quasi-elliptical rainbow lines lying in the neighboring channels. A detailed analysis of this evolution has shown that the lines within the rainbow pattern in the IP plane "interact" with each other, and that, as a result, their shapes change. These changes may lead to a qualitative change of the shape of the whole rainbow pattern. In the former above-mentioned case, the first and second rainbow lines "attract" each other and finally join. In the latter above-mentioned case, two parts of each quasi-rectangular rainbow line "repel" each other, and finally split.

But how do the proton trajectories in the above-discussed cases of proton-crystal interaction look like? How are the observed rainbows classified? These questions will be answered using five typical trajectories of transmitted protons for $E_0 = 2.0$ MeV calculated with the rainbow proton-atom interaction potential. The initial points of two of these trajectories, designated as trajectories I and II, are chosen to lie on the first rainbow line in the IP plane, being close to the channel center, and the initial points of the remaining three of them, designated as trajectories III, IV and V, on the second rainbow line in this plane, being close to the atomic string. The projections of these trajectories on the *xz* and *yz* planes are given in Figs. 4.5a, b, respectively. The inset of Fig. 4.5b depicts the initial points of the trajectories. Analysis has shown that trajectories I, II, and III include one deflection from a channel wall, and trajectories IV and V two deflections from the channel walls. We have analyzed the other trajectories of transmitted protons for $E_0 = 2.0$ MeV as well. The conclusion is that the trajectories of about 90% of all the transmitted protons include one deflection from a channel wall, and that the trajectories of the remaining about 10% of them, whose initial points lie very close to the atomic string, include two deflections from the walls. We have also concluded that all the trajectories with the initial points lying on the first rainbow line include one deflection from a channel wall, and that the trajectories with the initial points lying on the second rainbow line dominantly include two deflections from the channel walls. Therefore, it is evident that for $E_0 = 2.0$ MeV the rainbows in the TA plane corresponding to the first and second rainbows in the IP plane can be classified in the standard way as the primary and secondary rainbows, respectively. Analysis of the trajectories of transmitted protons for $E_0 = 1.5$, 1.0 and 0.7 MeV calculated with the rainbow interaction potential has given similar results, but, as expected, with the percentage of trajectories of all the transmitted protons that include one deflection from a channel wall being lower for a lower incident proton kinetic energy. On the basis of these results and taking into account the fact that the two rainbows join for a certain value of $E_0$ between 1.5 and 1.0 MeV, one may conclude that for $E_0 = 2.0$ and 1.5 MeV, the primary and secondary rainbows were

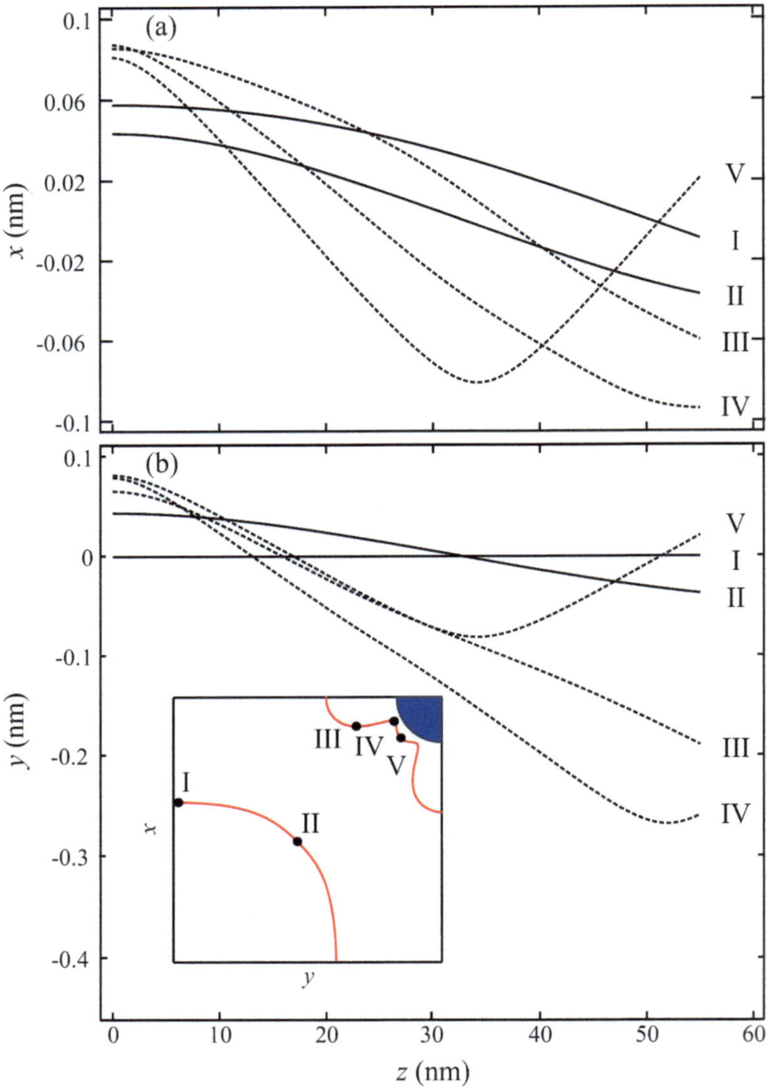


**Fig. 4.5** (**a**) Projections of five typical trajectories of 2.0 MeV protons transmitted through a 55-nm thick <100> Si crystal on the *xz* plane calculated for the rainbow ion-atom interaction potential [133]. (**b**) Associated projections of the proton trajectories on the *yz* plane [133]. *Inset*: Initial points of the proton trajectories

observed, while for $E_0 = 1.0$ and 0.7 MeV, the observed rainbows cannot be classified in the standard way.

We would like to emphasize that the above-described morphological method of modifying the $M^c(a_F)$ proton-atom interaction potential can be applied in each case of ion transmission through axial channels of a thin crystal where the rainbow pattern in the IP plane contains a line close to the channel center and a line close to the atomic string. Such a case can always be prepared by changing the values of $E_0$ and the crystal thickness, $L$. The adjusted rainbow pattern in the TA plane fully determines the associated angular distribution of transmitted ions, generated with the resulting rainbow interaction potential. This distribution ought to be compared with the corresponding distribution obtained in a high-resolution measurement. If they agree, one can conclude that a very accurate interaction potential has been extracted from this measurement. This interaction potential should also be very accurate for other values of $E_0$ and $L$ [51]. Unlike the ZBL, $\mathrm{M}(a_\mathrm{F})$, and $\mathrm{M}(a_\mathrm{TF})$ interaction potentials, it will be accurate across the whole channel. If this method is applied to different axial channels of a thin crystal, it will be possible to deduce an average rainbow interaction potential to accurately simulate the ion penetration through a randomly oriented crystal of the same atomic composition. This method has been also used to analyze the measurements of the Singapore group with 2.0 MeV protons transmitted through the <110> and <111> channels of a 55-nm thick (100) Si crystal [135]. The resulting parameters of the $\mathrm{M}^\mathrm{c}(a_\mathrm{F})$ interaction potential are given in Table 4.1.

Our opinion is that this morphological method may be applied in analogous ways in other cases where the rainbow effect occurs and plays an important role. These cases are reviewed in Chap. 1. That would lead to more accurate interaction potentials in the fields in question. It should be mentioned that a similar approach to atomic scattering from crystal surfaces has already produced important results [136, 137].

## 4.2 Inverse Transmission

In Subsect. 3.2.1, we have analyzed the rainbow effect in transmission of 10 MeV protons through a 100-nm thick <100> Au crystal. The crystal has been very thin. It has been shown that, in this process, the components of the proton impact parameter, $x_0$ and $y_0$, are single-valued and triple-valued functions of the components of the proton transmission angle, $\Theta_x$ and $\Theta_y$, on the inner side and outer side of the rainbow line in the TA plane, respectively. On the basis of this conclusion, one may consider the mapping of the TA plane to the IP plane, being inverse to the mapping of the IP plane to the TA plane, and say that the multiplicity of this mapping changes from 1 to 3 across the rainbow line in the TA plane. This means that it is possible to talk about the inverse proton transmission as the transmission determined by the functions $x_0(\Theta_x, \Theta_y)$ and $y_0(\Theta_x, \Theta_y)$.

In this section, we are going to explore the inverse transmission of 10 MeV protons through a 55-nm thick <100> Si crystal comprising the whole TA plane [138]. The crystal thickness, $L$, was chosen to be the same as in the experiments analyzed in Sect. 4.1, and the incident proton kinetic energy, $E_0$, was adjusted to make the crystal very thin – the reduced crystal thickness is $\Lambda = 0.07$. The incident proton velocity vector is parallel to the channel axis. The calculations were performed using a computer simulation method similar to the one described in Ref. [85] with the rainbow proton-atom interaction potential defined in the previous subsection. The differences between the two methods were in: (i) including the thermal vibrations of the crystal's atoms (see Sect.3.4), (ii) the numerical solving of the proton equations of motion in the transverse plane [128] instead of applying the model of crystal rainbows, and (iii) the construction of the surfaces defined by the functions $x_0(\Theta_x, \Theta_y)$ and $y_0(\Theta_x, \Theta_y)$ from the solutions of the equations of motions. This means that, as in the previous study, we disregarded the ion collisions with the crystal's electrons. The atomic strings of the crystal defining the channel intersected the lines $y = \pm x$. The number of atomic strings included in the calculations was 64, i.e., we took into account the strings lying on the four nearest square coordination lines.

Figure 4.6 depicts the surface representing the function $y_0(\Theta_x, \Theta_y)$ in a large region of the TA plane, comparable to the region determined by the critical angle for axial channeling, being $\psi_c = 2.37$ mrad. The surface representing the function $x_0(\Theta_x, \Theta_y)$ can be obtained by rotating the surface shown in the figure by 90° about the vertical axis passing through the origin. The values of the variables $x_0$ and $y_0$ were chosen uniformly from the uniform distribution within the region of the channel, and the incident number of protons was 976,208. It is evident that $y_0(-\Theta_x, -\Theta_y) = -y_0(\Theta_x, \Theta_y)$. Figure 4.7 shows the central part of the surface representing $y_0(\Theta_x, \Theta_y)$ viewed from above, i.e., along the $-y_0$ axis. One can clearly see the contour of this surface, whose projection to the $\Theta_x\Theta_y$ space, i.e., the TA plane, is a cusped square. Analysis shows that this projection coincides with the rainbow line in the TA plane (see Fig. 3.16).

It has been demonstrated in Subsect. 3.2.2 that the $^4X_9$ elementary catastrophe is the organizing center of the crystal rainbows in the cases of square, rectangular, centered rectangular, and hexagonal very thin crystals with one atomic string per

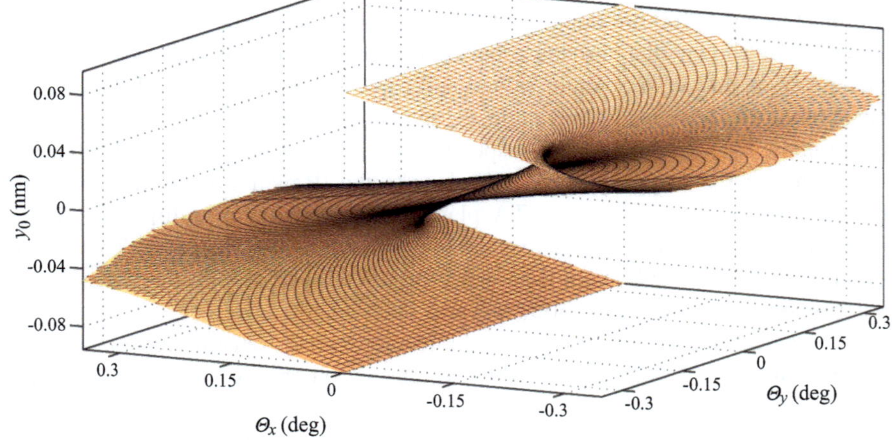


**Fig. 4.6** Surface representing the function $y_0(\Theta_x, \Theta_y)$ for 10 MeV protons transmitted through a 55-nm thick <100> Si crystal in a large region of the TA plane

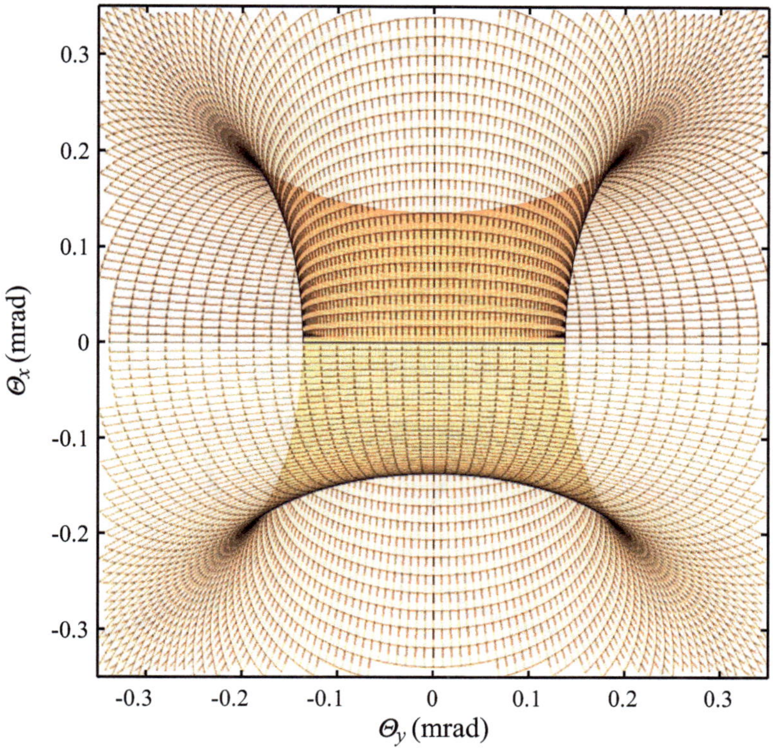


**Fig. 4.7** Central part of the surface representing the function $y_0(\Theta_x, \Theta_y)$ for 10 MeV protons transmitted through a 55-nm thick <100> Si crystal viewed along the $-y_0$ axis. The contour of the surface whose projection to the TA plane is a cusped square is the rainbow line in this plane

channel. As a result, we have obtained a universal, simple, and relatively accurate model proton-atom interaction potential. One of the considered cases has been the above-mentioned case of 10 MeV protons transmitted through a 100-nm thick <100> Au crystal. Here, we shall compare the surface defined by the above-depicted function $y_0(\Theta_x, \Theta_y)$, in the case of 10 MeV protons transmitted through a 55-nm thick <100> Si crystal, in the whole TA plane with the corresponding surface given by catastrophe theory. The appropriate form of the generating function of the $^4X_9$ catastrophe is its dual form, given by Eq. (3.26) with the positive first term. The values of the parameters of the corresponding catastrophic ion-crystal continuum interaction potential were obtained in the same way as in Subsect. 3.2.2. They are: $K = -10.46$, $\gamma = \delta = 0.402$, and $\rho_0 = 3.85$ a.u.

Figure 4.8 gives the lines representing the function $y_0(\Theta_x, \Theta_y)$ shown in Fig. 4.6 for $\Theta_x = 0.090$, 0.150, and 0.210 mrad. It is evident that, for the first value of $\Theta_x$, the multiplicity of the inverse proton transmission is 3 for $\Theta_y$ between −0.144 and 0.144 mrad, and it is 1 for $\Theta_y$ outside this interval. For the second value of $\Theta_x$, the multiplicity equals 3 for $\Theta_y$ between −0.163 and −0.109 mrad and between 0.109 and 0.163 mrad, and it equals 1 for $\Theta_y$ outside these intervals. Finally, for the third

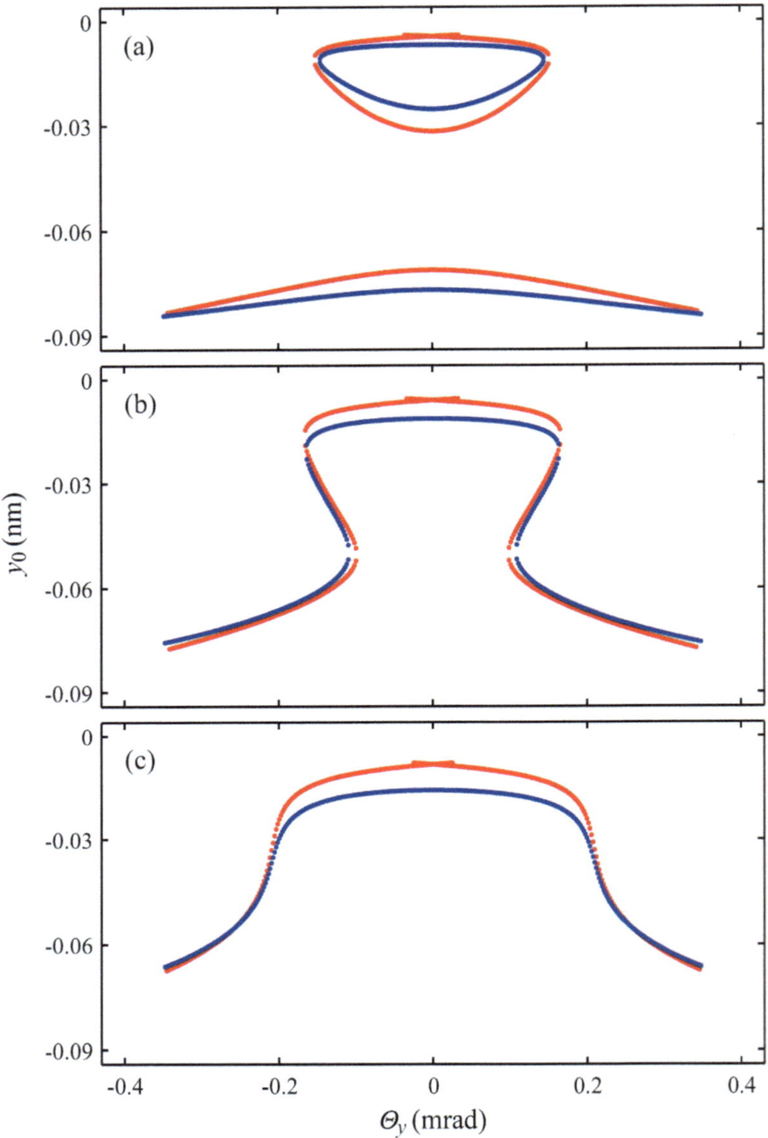


**Fig. 4.8** Lines representing the function $y_0(\Theta_x, \Theta_y)$ for 10 MeV protons transmitted through a 55-nm thick <100> Si crystal (*blue points*) and the corresponding lines calculated using the generating function of the ${}^4X_9$ elementary catastrophe (*red points*) for (**a**) $\Theta_x$=0.090, (**b**) $\Theta_x$=0.150, and (**c**) $\Theta_x$=0.210 mrad

value of $\Theta_x$, the multiplicity is equal to 1 for all the values of $\Theta_y$. Figure 4.9 shows the lines representing $y_0$ for $\Theta_y$=0.090, 0.150, and 0.210 mrad. One can see that, for the first value of $\Theta_y$, the multiplicity of the inverse proton transmission equals 3 for $\Theta_x$ between −0.144 and 0.144 mrad, and it equals 1 for $\Theta_x$ outside this interval. For

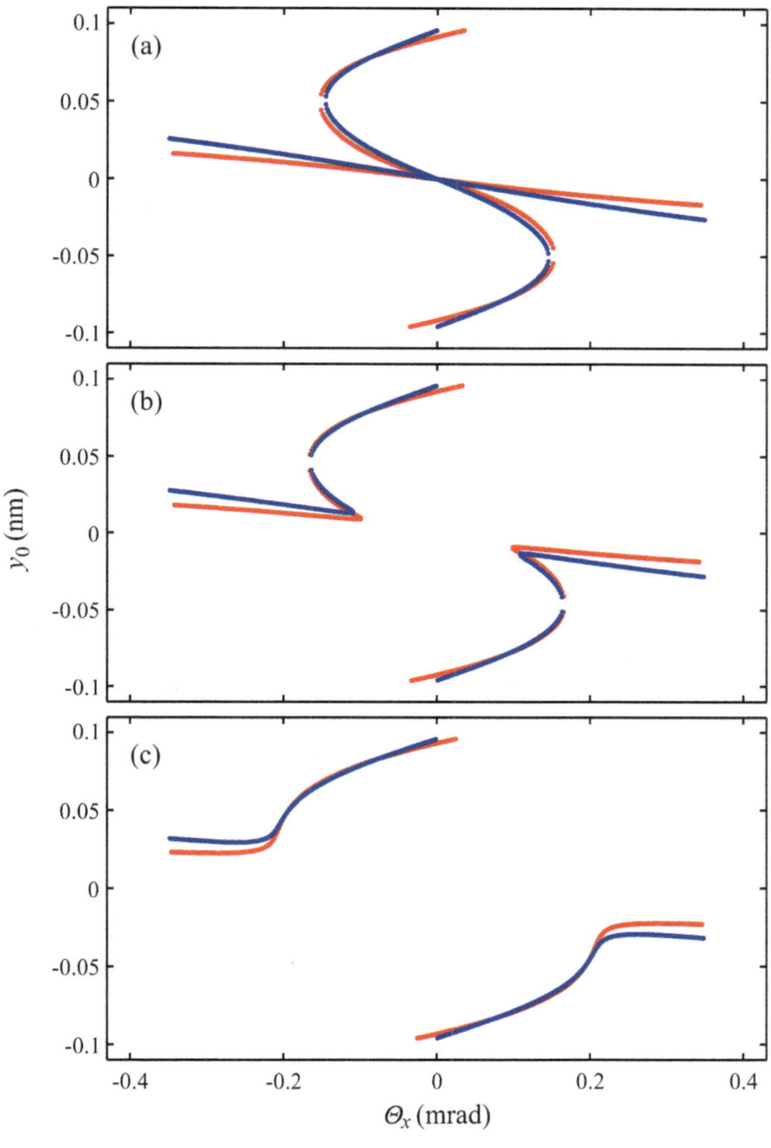


**Fig. 4.9** Lines representing the function $y_0(\Theta_x, \Theta_y)$ for 10 MeV protons transmitted through a 55-nm thick <100> Si crystal (*blue points*) and the corresponding lines calculated using the generating function of the $^4X_9$ elementary catastrophe (*red points*) for (**a**) $\Theta_y = 0.090$, (**b**) $\Theta_y = 0.150$, and (**c**) $\Theta_y = 0.210$ mrad

the second value of $\Theta_y$, the multiplicity is 3 for $\Theta_x$ between $-0.163$ and $-0.109$ mrad and between 0.109 and 0.163 mrad, and it is 1 for $\Theta_x$ outside these intervals. Finally, for the third value of $\Theta_y$, the multiplicity is equal to 1 for all the values of $\Theta_x$. All these conclusions are in agreement with the above-given statement that the multiplicity of the inverse transmission process changes from 1 to 3 across the rainbow line. When the multiplicity equals 3, the inverse process is partly probabilistic, meaning that, for a chosen pair of values of $\Theta_x$ and $\Theta_y$, the resulting pair of values of $x_0$ and $y_0$ appears with a probability less than 1. But when the probabilities of appearance of the three pairs of values of $x_0$ and $y_0$ corresponding to the chosen pair of values of $\Theta_x$ and $\Theta_y$ are summed, the result is 1. These two figures also contain the corresponding lines belonging to the equilibrium set of the $^4X_9$ elementary catastrophe. One can conclude that the agreement between the lines obtained by the computer simulation and catastrophe theory is excellent. The contour of the surface representing $y_0(\Theta_x, \Theta_y)$ corresponds to the catastrophic set of the $^4X_9$ elementary catastrophe, while the projection of this line to the TA plane corresponds to the bifurcation set of the catastrophe. This projection is the catastrophic model of the rainbow line in the TA plane.

All the above-presented facts, coming out of the analysis of the inverse process, demonstrate that the surface representing the function $x_0(\Theta_x, \Theta_y)$ or $y_0(\Theta_x, \Theta_y)$, whose shape depends on the employed ion-atom interaction potential, contains the basic information on the process of proton transmission in question. This surface may be referred to as the rainbow equilibrium surface. In an experiment, one measures the angular distribution of transmitted protons, whose "skeleton" is the projection of the contour of this surface, being the rainbow line in the TA plane. As it has been said above, the multiplicity of the inverse transmission process changes from 1 to 3 across this line, resulting in an abrupt change of the yield of transmitted protons across it, i.e., when passing from the dark side of the rainbow to its bright side. These statements make the explanation of the crystal rainbow effect in the spirit of catastrophe theory.

## 4.3 Superfocusing Effect

### *4.3.1 Axial Focusing*

It has been explained in Sect. 3.3 that a parallel ion beam entering an axial crystal channel aligned to its axis is almost parallel at the exit from the channel if the corresponding reduced crystal thickness is $\Lambda = n/2$, with $n = 1, 2, \ldots$ counting the rainbow cycles. This means that, for these values of $\Lambda$, the outgoing beam is focused around the origin of the TA plane. This is an effect of angular focusing, occurring at the end points of the rainbow cycles. As it has been said, it is called the effect of zero-degree focusing of channeled ions. But how does the incident parallel beam look like for $\Lambda = (n - 1)/2 + 1/4$, with $n = 1\ 2, \ldots$, i.e., at the middle points of the rainbow cycles?

It was found in the early days of studying ion propagation along axial crystal channels that, when the statistical equilibrium in the transverse plane is established, the yield of ions is maximal on the channel axis [66]. However, in the initial stage of the process, the yield of ions on the channel axis oscillates about the line close to the one corresponding to the statistical equilibrium. That effect was named the flux peaking effect in ion channeling [69, 139–151]. The main aim in all those studies was to measure the yield of ions that were backscattered from the foreign atoms inserted in the channels in order to determine their precise locations.

Demkov and Meyer [152] reopened the theoretical studying of the flux peaking effect before establishing the statistical equilibrium in the transverse plane. Since the effect was qualitatively different from the flux peaking effect after establishing the statistical equilibrium, they named it differently – the effect of superfocusing of channeled ions. It was a very inspiring study. Following our explorations of the zero-degree focusing effect, described in Sect. 3.3, the authors chose the proton beam of incident kinetic energy of $E_0 = 1$ MeV directed into the <100> channel of a Si crystal and studied its focusing around the origin in the final TP plane, i.e., its axial focusing, in the first rainbow cycle. They used the continuum approximation [66] and considered only the paraxial part of the channel. As a result, it was assumed that the proton-crystal continuum interaction potential was dominantly cylindrically symmetric and harmonic. Their main conclusions were: (i) the half-width of the superfocusing region, existing for $\Lambda$ about 0.25, went below 5 pm, (ii) its length was several tens of atomic layers, and (iii) the beam compression coefficient in it went up to several hundreds. The authors suggested that the superfocusing effect might be employed for subatomic microscopy and for increasing the luminosity of an ion beam collider.

Investigation of the superfocusing effect was continued by us [153]. We selected 2 MeV protons and the <100> channel of a Si crystal, and also concentrated on the effect occurring in the vicinity of the point $\Lambda = 0.250$. The calculations were done using the theory of crystal rainbows, presented in Sect. 3.4. The incident ion velocity vector is parallel to the channel axis. We adopt Moliére's ion-atom interaction potential defined by Eqs. (2.1), (2.2), and (2.4), and the continuum approximation [66], i.e., we use the corresponding continuum interaction potential of the ion and $i$th atomic string of the crystal, given by Eqs. (2.13), (2.14), (2.15), and (2.16). The frequency of proton oscillations around the channel axis, calculated from the second-order terms of the Taylor expansion of the continuum potential in the channel in the vicinity of the channel center, equals $f_h = 5.94 \times 10^{13}$ Hz. The atomic strings lie on the $x$ and $y$ axes (see Fig. 3.27). Their number is $M = 36$, i.e., we include the atomic strings lying on the three nearest square coordination lines. The thermal vibrations of the crystal's atoms are included, via the expression analogous to Eq. (3.13) [132], with the one-dimensional atomic thermal vibration amplitude being $\sigma_{th} = 0.00744$ nm [108, 109]. The proton collisions with the crystal's electrons are disregarded. The proton equations of motion in the transverse plane are solved numerically [128]. The components of the proton impact parameter vector are chosen randomly from the uniform distribution within the region of the channel taking into account the channel symmetry. The incident number of protons is 4,834,347. The Jacobian of $X$ and $Y$ [Eq. (3.37)] as well as the spatial rainbow lines in the IP plane [Eq. (3.38)] are determined numerically too [128].

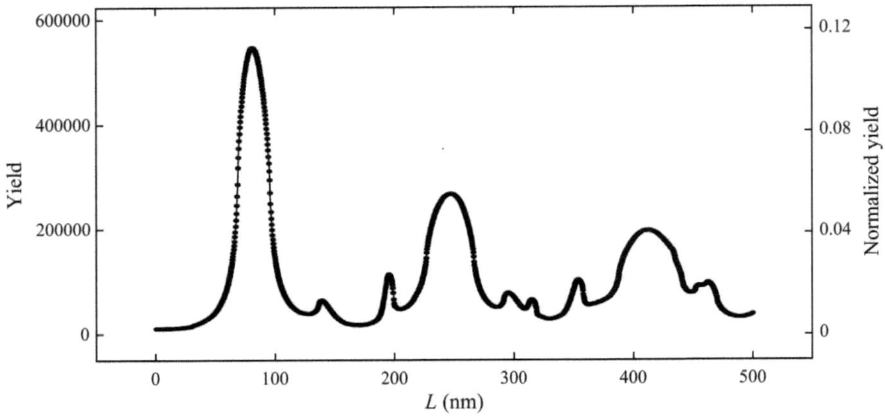


**Fig. 4.10** Axial and normalized axial yields of 2 MeV protons transmitted through the <100> channel of a Si crystal as functions of $L$ [153]

Figure 4.10 gives the yield of transmitted protons around the origin of the final TP plane, i.e., their axial yield, as a function of the crystal thickness, $L$. For the region in the final TP plane around the origin, we take the region in which the radial component of the proton position vector, $\rho$, is less or equal to $0.1a_0$, where $a_0 = 0.0529$ nm is the Bohr radius. This means that the radius of this region is taken to be much smaller than the average radius of a hydrogen atom in its ground state. This figure also shows the normalized yield of transmitted protons around the origin of the final TP plane, being the ratio of their axial yield and the incident number of protons, as a function of $L$. It is evident that this dependence is periodic and decays with $L$ [153]. In the chosen region of $L$, between 0 and 500 nm, it contains three pronounced maxima. We recognize that they occur due to the superfocusing effect, i.e., the corresponding points on the channel axis are the superfocusing points. The values of $\Lambda$ in question are 0.250, 0.750, and 1.250, while the corresponding values of $L$ are 83, 248, and 413 nm, respectively. Since the proton energy loss is neglected, the values of $\Lambda$ have been obtained by Eq. (3.16) with $v_{av}$ substituted with $v_0$, the magnitude of the incident ion velocity vector. Analysis has shown that the decay of the superfocusing effect with the increase of $L$ is caused mainly by the anharmonic component of the continuum potential in the channel, as in case of the zero-degree focusing effect (see Sect. 3.3). If we took into account the proton collisions with the crystal's electrons, the decay of the effect would be faster.

Figures 4.11a–c depict the distributions of transmitted protons in the final TP planes, i.e., their spatial distributions, for $\Lambda = 0.200$, 0.250, and 0.300. The spatial distribution for $\Lambda = 0.200$ contains a large cusped square ridge with a number of maxima along it. Four of these maxima are more pronounced than the others. They lie on the lines $Y = \pm X$, pointing between the atomic strings defining the channel. This spatial distribution also contains a net of small ridges. For $\Lambda = 0.250$, the spatial distribution contains a very much pronounced maximum at the origin and a

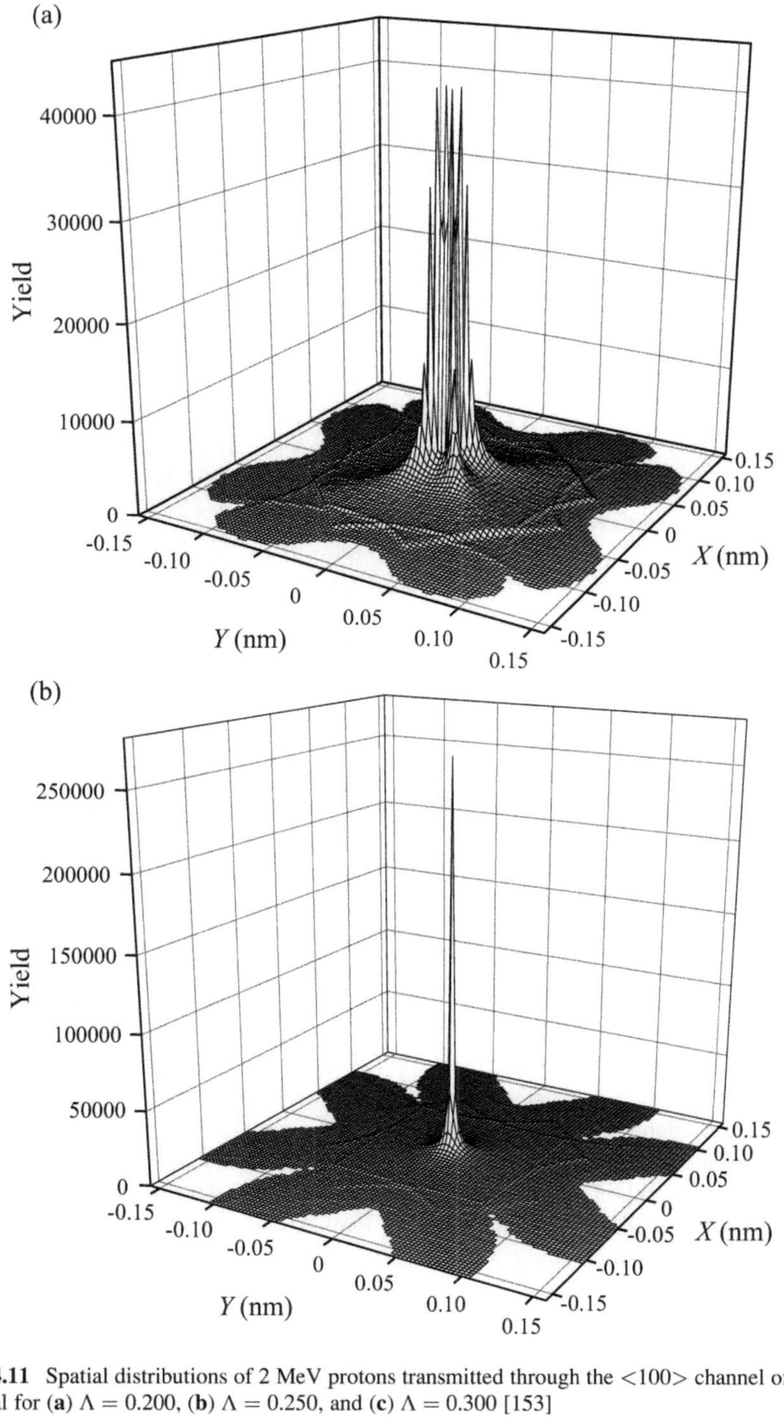


**Fig. 4.11** Spatial distributions of 2 MeV protons transmitted through the <100> channel of a Si crystal for (**a**) $\Lambda = 0.200$, (**b**) $\Lambda = 0.250$, and (**c**) $\Lambda = 0.300$ [153]

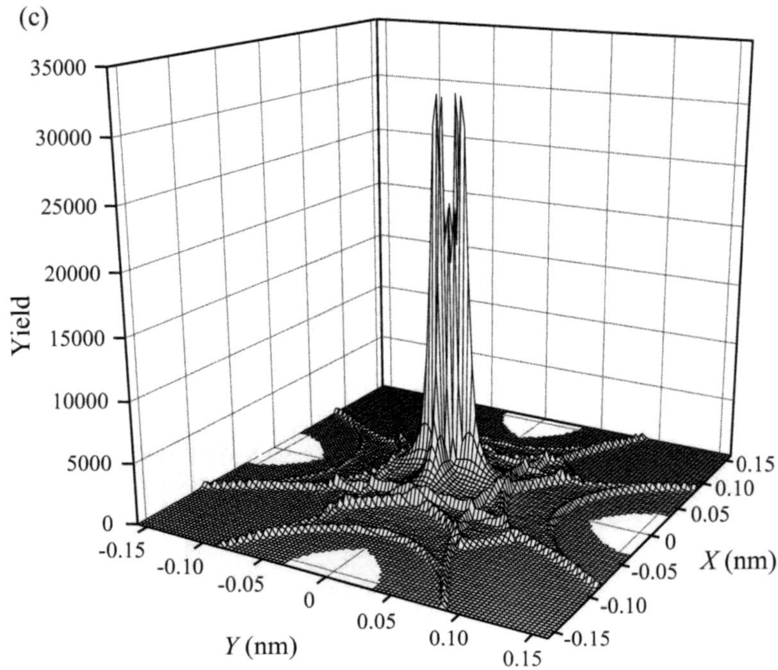


**Fig. 4.11** (continued)

net of very small ridges. The spatial distribution for $\Lambda = 0.300$ contains a large cusped square ridge with a number of maxima along it. Four of these maxima are more pronounced than the others. They lie on the $X$ and $Y$ axes, pointing toward rather than between the atomic strings defining the channel. This spatial distribution also contains a net of small ridges. It is important to note the differences between the scales of the yield axes in these three figures.

Analysis of the rainbow patterns in the final TP planes for $\Lambda = 0.200$, 0.250, and 0.300 has shown that, for $\Lambda = 0.200$, there is one cusped square rainbow line with the cusps directed between the atomic strings defining the channel; for $\Lambda = 0.250$, there is a rainbow point at the origin; and for $\Lambda = 0.300$, there is one cusped square rainbow line with the cusps directed toward rather than between the atomic strings defining the channel. Each of these patterns also contains a net of rainbow lines, which originate from the regions in the impact parameter plane close to the atomic strings defining the channel. Comparison of the spatial distributions and rainbow patterns for these three values of $\Lambda$ clearly demonstrates that: (i) all the ridges of the spatial distributions coincide with the rainbow lines, i.e., they occur due to the spatial crystal rainbow effect, (ii) the inner sides of the cusped square rainbow lines for $\Lambda = 0.200$ and 0.300 are the bright sides of the rainbows and their outer sides are the dark sides of the rainbows, and (iii) the very much pronounced maximum at the origin for $\Lambda = 0.250$ coincides with the rainbow point, i.e., it occurs due to the rainbow effect. The last statement means that the superfocusing effect is in fact a

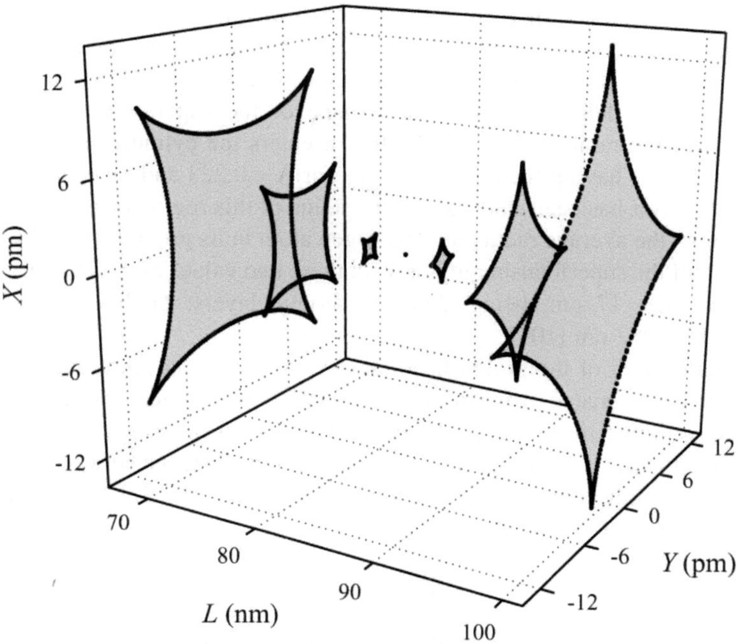


**Fig. 4.12** Evolution of the rainbow line in the final TP plane in the vicinity of the first superfocusing point for 2 MeV protons in the <100> channel of a Si crystal [153]. The chosen values of Λ are 0.200, 0.220, 0.240, 0.250, 0.260, 0.280, and 0.300. The bright sides of the rainbows are colored gray

spatial crystal rainbow effect. Let us now come back to Fig. 4.10 and say that the additional maxima it contains, besides the three maxima occurring due to the superfocusing effect, are caused by the rainbow lines passing through the region around the origin.

Figure 4.12 gives the rainbow patterns in the final TP planes in the vicinity of the superfocusing point in the first rainbow cycle for $\Lambda = 0.200$, 0.220, 0.240, 0.250, 0.260, 0.280, and 0.300. These patterns clearly demonstrate how the cusped square rainbow line with the cusps directed between the atomic strings defining the channel reduces to the rainbow point, being the first superfocusing point, and how the cusped square rainbow line with the cusps directed toward the atomic strings defining the channel emerges from the rainbow point. Thus, one can state that the superfocusing effect is a reduced spatial crystal rainbow effect, in which the rainbow line becomes a rainbow point. This means that the superfocusing effect cannot be fully explained with only the harmonic component of the continuum potential in the channel, as it was done by Demkov and Meyer [152]. The anharmonic component of the continuum interaction potential must be taken into account too, as it has been implicitly done in our calculations. The effect occurs as a result of the joint contribution of the harmonic and anharmonic components of the continuum interaction potential to the proton beam dynamics.

Since the inner side of each of the six rainbow lines shown in Fig. 4.12 is the bright side of the rainbow and its outer side the dark side of the rainbow, and, thus, the line is the envelope of the proton beam in the final TP plane [131], one can state that these lines define the extent and shape of the superfocusing region. The evolution of the rainbow line is such that it enters the cylindrical region around the channel axis having the radius of $0.1a_0$ for $\Lambda = 0.223$ and exits this region for $\Lambda = 0.275$. As it has been said above, the radius of this region is taken to be much smaller than the average radius of a hydrogen atom in its ground state. If we define the length of the superfocusing region with these two values of $\Lambda$, one can calculate that it is $\Delta L_f = 17$ nm, being about 32 atomic layers; in this case, 1 atomic layer = 0.543082 nm [107].

The half-width of the proton beam at the superfocusing point under consideration can be estimated using a simple quantum mechanical model [152]. The assumption is that in the paraxial part of the channel, in which $\rho$ is taken to be less or equal to $\rho_h = 33.9$ pm, being one quarter of the distance between the channel axis and each of the atomic strings defining the channel [92, 107], the proton-crystal continuum interaction potential is

$$V_h(\rho) = 2\pi^2 m_p f_h^2 \rho^2, \tag{4.1}$$

where $m_p$ is the proton mass. If the initial half-width of the proton wave packet equals $\rho_h$, the half-width of the proton wave packet at the superfocusing point is

$$\rho_s = \frac{\hbar}{2\pi m_p f_h \rho_h} = 5.0\text{pm}, \tag{4.2}$$

where $\hbar$ is the reduced Planck constant. Thus, taking that the half-width of the proton beam at the superfocusing point equals $\rho_s$, one finds that it is about 11 times smaller than $a_0$. This value justifies the above-chosen value of the radius of the region in the final TP plane around the origin ($0.1a_0$). Within this model, the compression coefficient of the proton beam at the superfocusing point is $\kappa_s = (\rho_h/\rho_s)^2 = 47$. It has been calculated that, when the anharmonic component of the continuum potential in the channel is included, $\kappa_s = 37$ [154].

### 4.3.2 *Rainbow Subatomic Microscopy*

Let us now verify in more detail the idea of Demkov and Mayer that the superfocusing effect may be employed for subatomic microscopy [152]. Following that idea, we have demonstrated in the previous subsection that the radius of a superfocused proton beam can go down to about 5.0 pm. This means that the approach has the potential for a breakthrough in the field of subatomic microscopy.

We shall analyze the superfocusing effect in the case of 68 MeV protons transmitted through the <100> channel of a Si crystal in the vicinity of the point

$\Lambda = 0.250$ for different values of the incident proton beam angle, i.e., crystal tilt angle [129]. The choice of this value of $E_0$ will be explained below. The problem will be solved applying the theory of crystal rainbows as in the previous subsection but with the following differences. The incident proton beam angle, $\varphi$, is varied between 0 and $\pm 0.30\psi_c$, where $\psi_c = 1.04$ mrad is the critical angle for axial channeling, defined by Eq. (2.12). The proton collisions with the crystal's electrons are taken into account. The corresponding energy loss rate is included in the calculations via the expression

$$-\frac{dE}{dz} = \frac{4\pi Z_1^2 e^4}{m_e v^2} n_e \ln \frac{2m_e v^2}{\hbar\omega_e}, \tag{4.3}$$

where $Z_1 = 1$ is the projectile atomic number, $m_e$ and $e$ are the electron mass and charge, respectively, $v$ is the magnitude of the projectile velocity vector, $n_e = \Delta V_{\mathrm{M}}^{\mathrm{th}}(x, y)/(4\pi)$ is the density of the target's electron gas averaged along the $z$ axis, $V_{\mathrm{M}}^{\mathrm{th}}$ is thermally averaged Moliére's projectile-target continuum interaction potential (see Sect. 3.4), $\Delta \equiv \partial_{xx} + \partial_{yy}$, and $\omega_e = (4\pi e^2 n_e/m_e)^{1/2}$ is the angular frequency of the electron gas oscillations induced by the projectile [51]. The rate of the corresponding average square uncertainty of the projectile channeling angle is taken into account in the calculations via the expression

$$\frac{d\sigma_e^2}{dz} = -\frac{m_e}{m_p^2 v^2}\frac{dE}{dz} \tag{4.4}$$

[51]. The incident number of projectiles is 4,352,720.

Figure 4.13 shows the yields of transmitted protons around the origin in the final TP plane as functions of $\Lambda$ and $L$ in the first rainbow cycle for $\varphi = 0$ without taking into account the proton collisions with the crystal's electrons. For the region in the final TP plane around the origin, we take the region in which $\rho$ is below $a_1 = 0.01a_0 = 0.5$ pm. These dependences contain a maximum at $\Lambda_m = 0.250$, corresponding to $L_m = 482$ nm, being the superfocusing point. Hence, one concludes that this maximum occurs due to the superfocusing effect, which is explained by the dominance of the harmonic component of the proton-crystal continuum interaction potential over its anharmonic component in the region close to the channel axis [152, 153]. Its width, being a measure of the longitudinal extent of the superfocusing region, is $\Delta\Lambda_f = 0.018$ or $\Delta L_f = 35$ nm, being about 64 atomic layers [107]. The figure also gives the axial yields of transmitted protons for $a_1 = 0.5$ pm with the proton collisions with the crystal's electrons taken into account. Now, the superfocusing maximum is much weaker and broader than the corresponding maximum obtained without taking into account these collisions. It is located at $\Lambda_m = 0.235$, corresponding to $L_m = 453$ nm, and its width is $\Delta\Lambda_f = 0.088$ or $\Delta L_f = 170$ nm, being about 312 atomic layers [107]. The weakening and broadening of the superfocusing maximum are attributed to the spreading of the spatial distribution of transmitted protons caused by the proton collisions with the crystal's electrons. Its shift from the superfocusing point, being $\Delta\Lambda_m = 0.015$ or

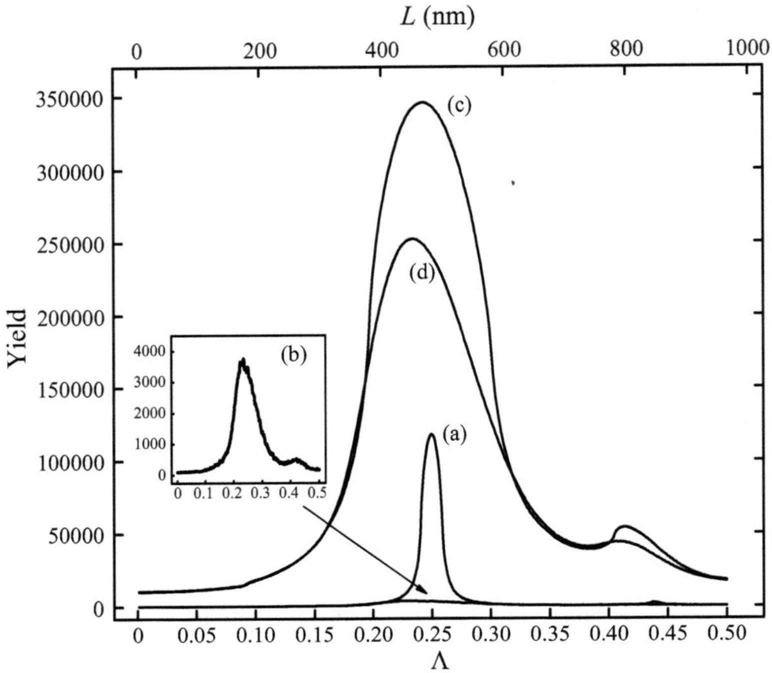


**Fig. 4.13** Axial yields of 68 MeV protons transmitted through the <100> channel of a Si crystal for (**a**) $a_1 = 0.5$ pm without including the proton collisions with the crystal's electrons, (**b**) $a_1 = 0.5$ pm with these collisions, (**c**) $a_1 = 7.5$ pm without the collisions, and (**d**) $a_1 = 7.5$ pm with the collisions as functions of $\Lambda$ and $L$ in the first rainbow cycle for $\varphi = 0$ [129]

$\Delta L_m = 29$ nm, can be explained if the evolution of the rainbow line in the final TP plane in the vicinity of the superfocusing point described in the previous subsection is taken into account. It is caused by the difference between the evolution of the rainbow line after the superfocusing point, for $\Lambda$ above 0.250, and the time-reversed evolution of the rainbow line before the superfocusing point, for $\Lambda$ below 0.250. In the former case, the rainbow line moves away from the origin faster than in the latter case, resulting in a decreased axial proton yield in the former case or an increased yield in the latter case. This difference becomes observable due to the spreading of the spatial distribution. It is evident that the weakening and broadening of the superfocusing maximum when one takes into account the proton collisions with the crystal's electrons are not negligible. If $E_0$ were chosen to be 2 MeV, as in the previous subsection, this would be even more pronounced. That is why, in this study, $E_0 = 68$ MeV. A series of relevant experiments were conducted with this value of $E_0$[155, 156].

Besides, Fig. 4.13 gives the axial yields of transmitted protons for $a_1$=7.5 pm without taking into account the proton collisions with the crystal's electrons as functions of $\Lambda$ and $L$. The choice of this value of $a_1$ will be explained later. In this case, the superfocusing maximum is located at $\Lambda_m = 0.243$, corresponding to

$L_m = 468$ nm, and its width is $\Delta\Lambda_f = 0.105$ or $\Delta L_f = 202$ nm, being about 373 atomic layers [107]. The small shift of the superfocusing maximum from the superfocusing point, being $\Delta\Lambda_m = 0.002$ or $\Delta L_m = 3.9$ nm, is attributed to the difference between the evolution of the rainbow line in the final TP plane after the superfocusing point and the time-reversed evolution of the rainbow line before the superfocusing point, which becomes observable due to the larger value of $a_1$. This figure also shows the axial yields of transmitted protons for $a_1 = 7.5$ pm with the proton collisions with the crystal's electrons taken into account. As in the case of $a_1 = 0.5$ pm, the superfocusing maximum is weaker and broader than the corresponding maximum obtained without including these collisions. It is located at $\Lambda_m = 0.234$, corresponding to $L_m = 451$ nm, and its width is $\Delta\Lambda_f = 0.114$ or $\Delta L_f = 220$ nm, being about 405 atomic layers [107]. Again, the weakening and broadening of the superfocusing maximum are attributed to the spreading of the spatial distribution of transmitted protons caused by the proton collisions with the crystal's electrons. Its larger shift from the superfocusing point, being $\Delta\Lambda_m = 0.016$ or $\Delta L_m = 31$ nm, is again explained by the difference between the evolution of the rainbow line after the superfocusing point and the time-reversed evolution of the rainbow line before the superfocusing point, which becomes more observable due to the spreading of the spatial distribution.

Figure 4.14a gives the spatial distributions of transmitted protons along the $X$ axis for $\varphi = 0$, $\pm 0.10\psi_c$, $\pm 0.20\psi_c$, and $\pm 0.30\psi_c$, and $\Lambda = 0.250$ with the proton collisions with the crystal's electrons taken into account. Each of these distributions contains one maximum, appearing due to the superfocusing effect. The width of the maximum for $\varphi = 0$, lying at the origin, is $\Delta X_m = 15.0$ pm. This width represents the proton beam diameter for $\varphi = 0$ and $\Lambda = 0.250$. The calculations have shown that for $\Lambda$ above 0.250, $\Delta X_m$ increases abruptly; for $\Lambda$ between 0.250 and 0.225, it practically stays the same; and for $\Lambda$ below 0.225, it increases abruptly too. This is illustrated in Fig. 4.14b. Thus, one can call the part of the superfocusing region between $\Lambda = 0.225$ and 0.250 the beam neck. Its length is $\Delta\Lambda_n = 0.025$ or $\Delta L_n = 48$ nm. Accordingly, we have decided to adopt that within the beam neck the value of its diameter is equal to its value for $\Lambda = 0.250$ (15.0 pm). This has been the reason for choosing the larger value of $a_1$ (7.5 pm) in the analysis of the axial yields of transmitted protons as functions of $\Lambda$ and $L$, the results of which are given in Fig. 4.13.

As one would expect, for the positive values of $\varphi$, the maxima of the spatial distributions of transmitted protons given in Fig. 4.14a lie on the positive part of the $X$ axis, and, for the negative values of $\varphi$, on the negative part of the $X$ axis. The larger the modulus of $\varphi$, the larger is the displacement of the maximum from the origin. Also, when the modulus of $\varphi$ is larger, the maximum is weaker. The positions of the maxima for $\varphi = \pm 0.10\psi_c$, $\pm 0.20\psi_c$, and $\pm 0.30\psi_c$ are $\pm 31.8$, $\pm$ 60.3, and $\pm 76.2$ pm, respectively, while their heights are 97.3, 84.9, and 49.4% of the height of the maximum for $\varphi = 0$, respectively. Therefore, one can say that the maxima for $\varphi = \pm 0.10\psi_c$ and $\pm 0.20\psi_c$, whose displacements from the origin are below $1.3a_0$ and above $-1.3a_0$, respectively, are strong, as the maximum for $\varphi = 0$. On the other hand, the maxima for $\varphi = \pm 0.30\psi_c$, whose displacements from the

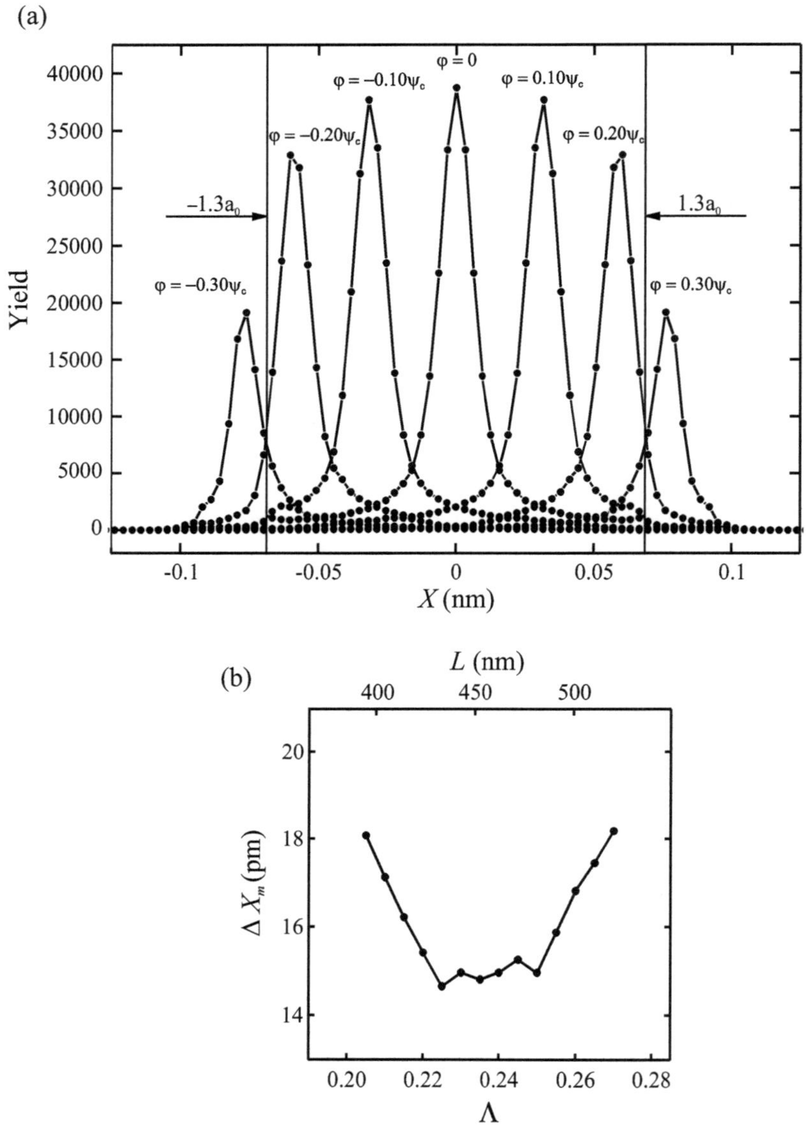


**Fig. 4.14** (**a**) Yields of 68 MeV protons transmitted through the <100> channel of a Si crystal along the $X$ axis in the final TP plane for $\varphi = 0$, $\pm 0.10\psi_c$, $\pm 0.20\psi_c$, and $\pm 0.30\psi_c$, and $\Lambda = 0.250$, with the proton collisions with the crystal's electrons taken into account [129]. (**b**) Widths of the spatial distributions of transmitted protons along the $X$ axis in the final TP plane as functions of $\Lambda$ and $L$ in the vicinity of the point $\Lambda = 0.250$ for $\varphi = 0$ with these collisions [129]

origin are above $1.3a_0$ and below $-1.3a_0$, respectively, are weak. This suggests that, by varying $\varphi$ from 0 to $\pm 0.20\psi_c$, one can probe the interior of a foreign atom inserted in the channel at a depth around $\Lambda = 0.250$, i.e., in the vicinity of the superfocusing point [152]. The weakening of the superfocusing effect for

$\varphi = \pm 0.30\psi_c$ can be attributed to the fact that, in this case, the proton beam is considerably displaced from the channel axis toward the upper or lower channel wall, respectively, and moves through the region of the channel in which the anharmonic component of the continuum potential and the proton collisions with the crystal's electrons are more pronounced than in the region close to the channel axis [153]. It should be noted that the spatial distributions given in the figure have been calculated for $\Lambda = 0.250$ in spite of the fact that the maximum of the dependence (d) shown in Fig. 4.13 lies at $\Lambda_m = 0.234$. The reasons for such a choice have been the facts that the shift of that maximum from the superfocusing point, being $\Delta\Lambda_m = 0.016$, was within the beam neck, extending between $\Lambda = 0.225$ and 0.250, and that the point $\Lambda = 0.250$ is the superfocusing point.

The spatial distributions of transmitted protons in the two-dimensional representation for $\varphi = 0$, $\pm 0.10\psi_c$, $\pm 0.20\psi_c$, and $\pm 0.30\psi_c$, and $\Lambda = 0.250$ are shown in Fig. 4.15. It is evident that each of the distributions has one maximum. The shapes of the areas of the distributions for $\varphi = 0$ and $\pm 0.10\psi_c$ representing the constant

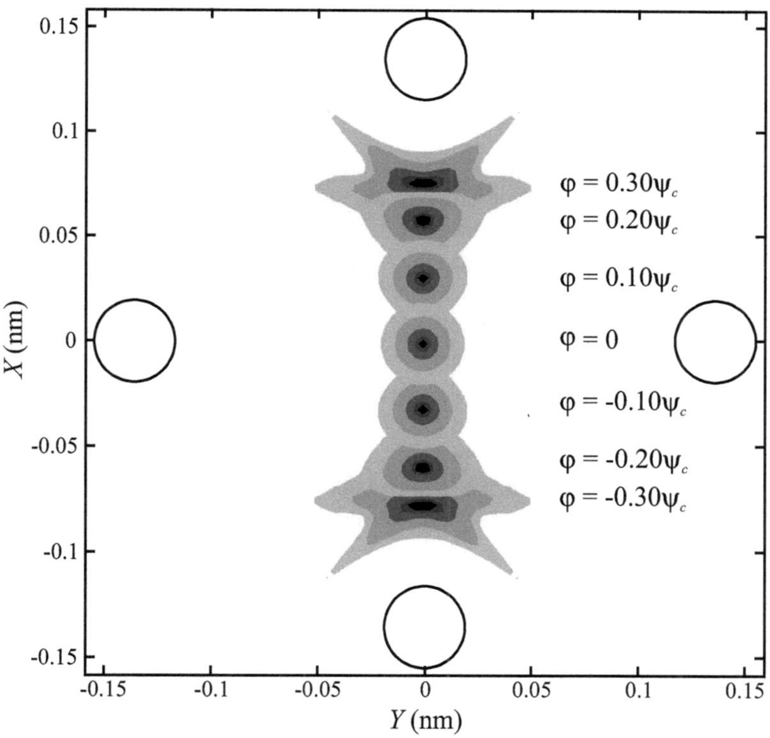


**Fig. 4.15** Yields of 68 MeV protons transmitted through the <100> channel of a Si crystal in the final TP plane in the two-dimensional representation for $\varphi = 0$, $\pm 0.10\psi_c$, $\pm 0.20\psi_c$, and $\pm 0.30\psi_c$, and $\Lambda = 0.250$ with the proton collisions with the crystal's electrons taken into account [129]. The areas in which the yields of transmitted protons are above 80, 50, 20, 15, 10, and 5% of the maximal yield are designated by the decreasing tones of *gray color*. Four *open circles* represent the atomic strings defining the channel

proton yields are circular, while the shapes of the areas of the distributions for $\varphi=\pm0.20\psi_c$ are quasi-ellipsoidal. The widths of the distributions along the $X$ and $Y$ axes for $\varphi=0$, $\pm0.10\psi_c$, and $\pm0.20\psi_c$, $\Delta X_m$ and $\Delta Y_m$, are 15 and 15 pm, 15 and 16 pm, and 14 and 20 pm, respectively. For $\varphi=\pm0.30\psi_c$, the distributions become broader, and the shapes of their central areas representing the constant proton yields become quasi-rectangular. It is clear that, for these values of $\varphi$, the superfocusing effect is attenuated and dissolved. Thus, as it has been suggested above, the possibility to use the proton beam for probing the interior of a foreign atom inserted in the channel in the vicinity of the superfocusing point exists only for the values of the modulus of $\varphi$ below $0.20\psi_c$. Figures 4.16a, b give the spatial distributions of transmitted protons in the three-dimensional representation for $\varphi=0.10\psi_c$ and $0.20\psi_c$, respectively.

In the analysis presented in Figs. 4.14, 4.15, and 4.16, the direction of the incident proton beam is changed along the $X$ axis, i.e., toward the atomic strings defining the channel. We have also explored the spatial distributions of transmitted protons for $\Lambda=0.250$ and the direction of the incident beam changing along line $Y=X$, i.e., between the atomic strings defining the channel. The obtained results are similar to those presented in Figs. 4.14, 4.15, and 4.16.

We have also investigated the dependences of the yields of the transmitted protons around the centers of their spatial distributions on $\Lambda$ in the region between 0.200 and 0.300 for $\varphi=0.10\psi_c$ and $0.20\psi_c$ with the proton collisions with the crystal's electrons included. For the region in the final TP plane around the center, we have taken the circular region having the radius equal to $a_1=7.5$ pm, as in cases of the dependences (c) and (d) shown in Fig. 4.13. This is in accordance with the facts demonstrated in Fig. 4.15 that the shapes of the areas of the distributions for $\Lambda=0.250$, and $\varphi=0$ and $0.10\psi_c$ representing the constant proton yields, are circular while the shapes of the areas of the distributions for the same value of $\Lambda$ and $\varphi=0.20\psi_c$ are quasi-ellipsoidal. For $\varphi=0.20\psi_c$, the circular region has approximated the quasi-ellipsoidal region. The obtained dependences are similar to the dependence (d) displayed in Fig. 4.13, obtained for $\varphi=0$.

Figure 4.17 gives the rainbow patterns in the final TP plane for $\varphi=0$, $0.10\psi_c$ and $0.20\psi_c$, and $\Lambda=0.250$. As it has been explained in Sect. 3.4, they are obtained without taking into account the proton collision with the crystal's electrons. For $\varphi=0$, the rainbow line reduces to the rainbow point, being the superfocusing point [153]. The rainbow patterns for $\varphi=-0.10\psi_c$ and $-0.20\psi_c$ and the same value of $\Lambda$ lie in the final TP plane symmetrically with respect to the $Y$ axis to the corresponding patterns shown in the figure. One can see that each of the rainbow patterns for $\varphi>0$ consists of two cusped triangular rainbow lines, each of which has two cusps directed between the atomic strings defining the channel and the third cusp directed along the $Y$ axis and toward the third cusp of the other line. The inner sides of these lines are the bright sides of the rainbows, while their outer sides are the dark sides of the rainbows. As $\varphi$ increases, the rainbow pattern shifts along the $X$ axis. Also, with the increase of $\varphi$, the two rainbows become larger and more separated from each other. Comparison of this figure with Figs. 4.15 and 4.16 demonstrates that, in spite of the fact that each rainbow pattern is composed of two

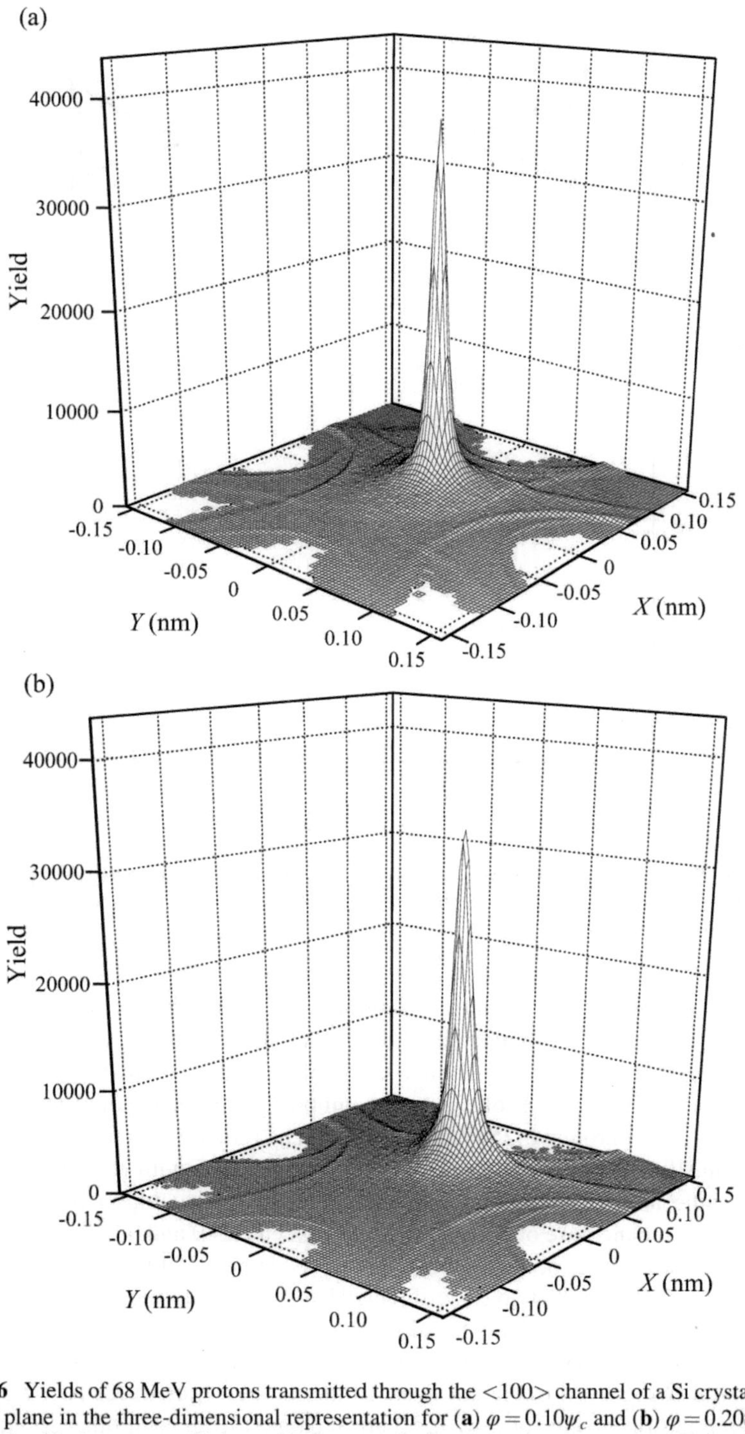


**Fig. 4.16** Yields of 68 MeV protons transmitted through the <100> channel of a Si crystal in the final TP plane in the three-dimensional representation for (**a**) $\varphi = 0.10\psi_c$ and (**b**) $\varphi = 0.20\psi_c$, and $\Lambda = 0.250$, with the proton collisions with the crystal's electrons taken into account (Adapted from Ref. [129])

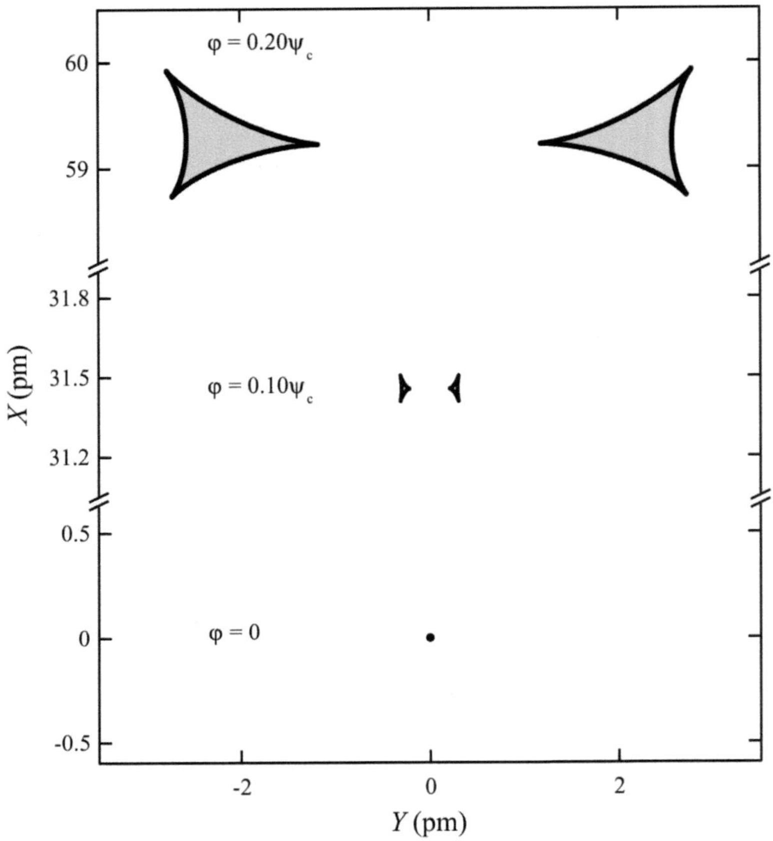


**Fig. 4.17** Evolution of the rainbow pattern in the final TP plane with an increase of $\varphi$, taking the values 0, $0.10\psi_c$, and $0.20\psi_c$, for $\Lambda = 0.250$. The bright sides of the rainbows are colored gray (Adapted from Ref. [129])

parts, the corresponding spatial distribution of transmitted protons contains only one maximum. This is explained by the fact that, for each of the considered values of $\varphi$, the rainbow pattern lies within a circle of the radius as small as 2.5 pm.

Let us also analyze the evolution of the rainbow pattern in the final TP plane with the increase of $\Lambda$ around 0.250 for $\varphi = 0.10\psi_c$. The rainbow patterns for $\Lambda = 0.200$, 0.220, and 0.240 consist of a cusped rectangular line with the cusps directed between the atomic strings defining the channel. The evolution of the rainbow pattern with the increase of $\Lambda$ in the region between 0.240 and 0.260 is displayed in Fig. 4.18. For $\Lambda = 0.245$, the rainbow pattern is composed of two cusped triangular lines, each of which with two cusps directed between the atomic strings defining the channel and the third cusp directed along the $Y$ axis and toward the third cusp of the other line. The rainbow pattern for $\Lambda = 0.250$, which is also shown in Fig. 4.17, consists of two cusped triangular lines, as the rainbow pattern for $\Lambda = 0.245$. The two rainbows occurring for $\Lambda = 0.250$ are separated from each other less than the

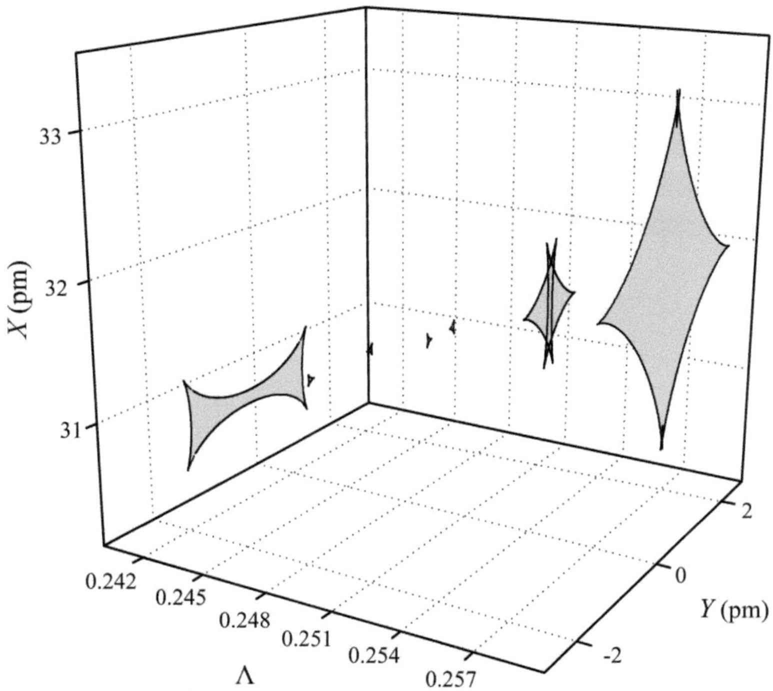


**Fig. 4.18** Evolution of the rainbow pattern in the final TP plane with an increase of Λ in the vicinity of the first superfocusing point, taking the values 0.240, 0.245, 0.250, 0.255, and 0.260, for $\varphi = 0.10\psi_c$. The bright sides of the rainbows are *colored gray* (Adapted from Ref. [129])

two rainbows occurring for $\Lambda = 0.245$. For $\Lambda = 0.255$, the rainbow pattern contains two cusped triangular lines, each of which with two cusps directed between the atomic strings defining the channel and the third cusp directed along the *Y* axis and away from the third cusp of the other line. These two rainbows are larger than the two rainbows occurring for $\Lambda = 0.250$, and they overlap. The rainbow pattern for $\Lambda = 0.260$ consists of a deltoidal line with the two joints of its sides having the form of a butterfly directed along the *X* axis and the other two joints of its sides having the form of a cusp directed along the *Y* axis. The rainbow patterns for $\Lambda = 0.280$ and 0.300, which are not shown in Fig. 4.18, contain a cusped deltoidal line with the cusps directed toward rather than between the atomic strings defining the channel. As in Fig. 4.17, the inner sides of all these lines are the bright sides of the rainbows, while their outer sides are the dark sides of the rainbows.

It is interesting to note that the sequences of forms displayed in Figs. 4.12, 4.17, and 4.18 coincide qualitatively with the unfoldings of the $^4X_9$ elementary catastrophe that were analyzed in detail by Nye in the context of light reflection [12, 18]. This coincidence is in agreement with our earlier finding, described in Subsect. 3.2.2, that the $^4X_9$ catastrophe acts as the organizing center of crystal rainbows appearing with protons and square, rectangular, centered rectangular, and

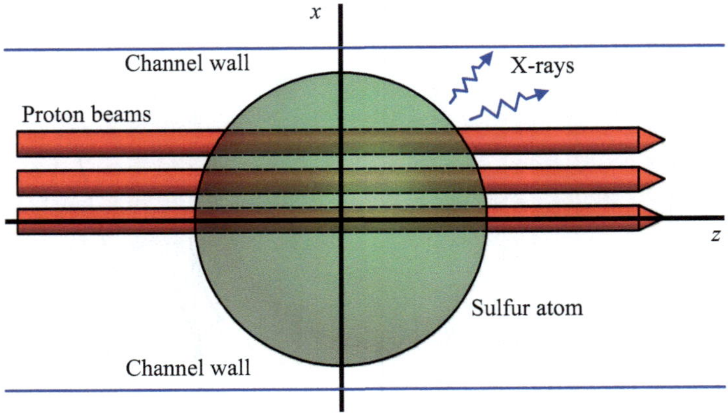


**Fig. 4.19** Illustration of the interactions of three 68 MeV proton beams, for $\varphi = 0$ , $0.10\psi_c$, and $0.20\psi_c$, with the inner-shell electrons of a sulfur atom inserted in the <100> channel of a Si crystal at the first superfocusing point resulting in the emissions of characteristic X-rays (PIXE) [129]

hexagonal very thin crystals with one atomic string per channel. We think that this catastrophe could be used for a simple and qualitatively accurate modeling of the changes of the rainbow pattern with $\varphi$ and $\Lambda$ around the superfocusing point.

We are now going to consider a foreign atom localized in the channel in the vicinity of the superfocusing point and exposed to the proton beam. One of the results of the proton-foreign atom interaction can be an inner-shell ionization of the atom resulting in the emission of a characteristic X-ray, which is referred to as proton-induced X-ray emission (PIXE) [155, 156]. This is illustrated in Fig. 4.19. The distance between the opposite channel walls is 271.5 pm and the radius of the foreign atom is $R_a = 100$ pm, corresponding to a sulfur atom [157]. Three beams are shown – for $\varphi = 0$, $0.10\psi_c$, and $0.20\psi_c$. Since, for $\varphi = 0$, $R_a$ is much smaller than the length of the beam neck, being $\Delta L_n = 48$ nm, we take that the corresponding beam diameter is constant in the interaction region and that it is equal to its diameter at the superfocusing point, being $\Delta X_m = 15.0$ pm. The situation is the same for $\varphi = 0.10\psi_c$ and $0.20\psi_c$. For $\varphi = 0$, the protons propagate through the foreign atom along its diameter coinciding with the $z$ axis, and for $\varphi = 0.10\psi_c$ and $0.20\psi_c$, they propagate through the atom along its chords being practically parallel to the $z$ axis. The larger the value of $\varphi$, the smaller is the length of the chord. This simple geometrical analysis demonstrates the possibility for measuring the differential cross-section for PIXE as functions of the components of the proton impact parameter vector within the foreign atom. Thus, one could measure the electron density within the foreign atom. As it has been said above, such a probing of the interior of the foreign atom has been named the rainbow subatomic microscopy [153]. It is a measurement technique with the picometer resolution.

The realization of the idea of rainbow subatomic microscopy depends crucially on the possibility to localize the foreign atom in the vicinity of the superfocusing point, i.e., to make it lie within the proton beam neck. The displacement of the foreign atom from the beam neck can be corrected by varying $E_0$ upward or downward to move the neck forward or backward, respectively, and make the atom lie within the neck. However, the effect of thermal vibrations of the foreign atom makes its position uncertain. Since $\sigma_{\mathrm{th}}$ is comparable to the average beam radius [108, 109], one should cool the crystal to a temperature well below room temperature to make the effect less pronounced. But, even at the temperature of $T = 0$ K, the quantum vibrations of the foreign atom would make the uncertainty of its position comparable to the characteristic dimension of its K-shell. Using the Debey approximate treatment of the thermal vibrations [51] and an estimated value of the Debey temperature of the foreign atom, being 300 K [158], one can calculate that for $T = 0$ K, the one-dimensional thermal vibration amplitude of the foreign atom is 6.2 pm. On the other hand, the characteristic dimension of the K-shell of the foreign atom is 5.1 pm, and the characteristic dimensions of its L-shell and M-shell are about 25 and 100 pm, respectively [159]. This means that the effect of thermal vibrations is an obstacle for measuring the proton-induced emission of X-rays from the K-shell of the foreign atom, but that it does not prevent one to measure precisely the emissions from its L-shell and M-shell.

The resolution of the rainbow subatomic microscopy is determined by the ratio of $R_a$ and the average proton beam radius. In the example we consider, the values of this ratio are 13.3, 12.9, and 12.5 for the three values of $\varphi$, respectively. Thus, one can choose five values of $\varphi$: 0, $\pm 0.10\psi_c$, and $\pm 0.20\psi_c$, and obtain five well-separated measurement points. If the differential cross-section for PIXE is measured, the final result of the experiment would be the transverse projection of the electron density within the foreign atom. In this case, the sensitivity of the microscopy is determined by the way the differential cross-section depends on the electron density.

However, in order to observe experimentally the superfocusing effect, one must have an array of foreign atoms in the channels with the same transverse positions relative to the channel axes and the same longitudinal positions relative to the entrance plane of the crystal. Since the proton beam radius is considerably below $R_a$, the transverse positions of the foreign atoms must coincide with each other. However, the longitudinal distances of the foreign atoms from the superfocusing point do not have to coincide with each other. Instead, they must be in the region between $-\Delta L_n/2$ and $\Delta L_n/2$. One way of creating the required array of foreign atoms would be by the technique of ion implantation. The foreign atoms would be implanted in a Si crystal to form a buried single-crystalline layer with the foreign atoms at the front interface of the obtained three-layer structure facing the channels of the Si crystal [160, 161]. After the implantation, one would have to anneal the obtained structure to make the transverse positions of the foreign atoms within the channels at the interface coincide with each other. Such an interface would contain the required array of foreign atoms. The other way of creating the required array of foreign atoms would be by the technique of epitaxial growth. The idea is to grow a single-crystalline layer on a Si crystal with the foreign atoms at the interface of the obtained two-layer structure facing the channels of the Si crystal [149]. This would have to be followed by etching the Si crystal to adjust its thickness.

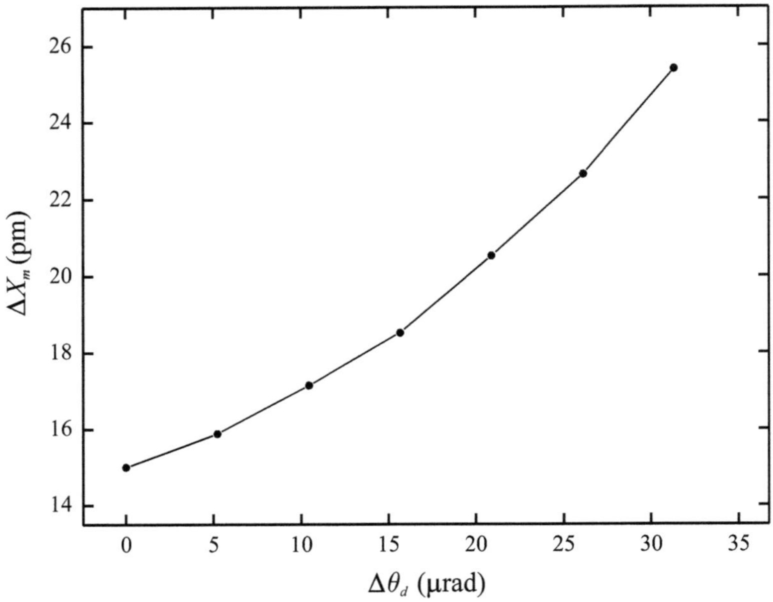


**Fig. 4.20** Width of the spatial distribution of 68 MeV protons transmitted through the <100> channel of Si crystal along the $X$ axis in the final TP plane as a function of $\Delta\theta_{\mathrm{d}}$ for $\varphi = 0$ and $\Lambda = 0.250$ with the proton collisions with the crystal's electrons taken into account [129]

In addition, in a superfocusing experiment, one must pay special attention to the incident proton beam divergence, $\Delta\theta_d$. Figure 4.20 gives the dependence of $\Delta X_m$ on $\Delta\theta_d$ for $\varphi = 0$ and $\Lambda = 0.250$ with the proton collisions with the crystal's electrons taken into account. It is clear that for $\Delta\theta_d$ below about 20 μrad, $\Delta X_m$ is below about 20 pm. The analysis has shown that the same is true for $\varphi = 0.10\psi_c$ and $0.20\psi_c$. Taking into account the fact, which can be seen in Fig. 4.14a, that the average distance between the superfocusing maxima for $\varphi = 0$ and $0.10\psi_c$ and between the ones for $\varphi = 0.10\psi_c$ and $0.20\psi_c$ is 30.2 pm, one can conclude that for $\Delta\theta_d$ below about 20 μrad, these superfocusing maxima would be clearly separated. Such a small value of $\Delta\theta_d$ has already been attained experimentally [162]. For example, the beam collimation system consisting of two apertures of the diameters of 100 μm placed at the distance of 5 m from each other would guarantee that $\Delta\theta_d$ would be below 20 μrad.

### 4.3.3 Measurement of the Superfocusing Effect

The Singapore group conducted a sequence of high-resolution measurements with 1.00, 0.90, 0.85, 0.75, 0.70, and 0.65 MeV focused proton microbeams transmitted through a 55-nm thick <100> Si crystal, and it was analyzed by them and us

[134]. As in their previous measurements [45–48, 130, 133], the angular distributions of transmitted protons were recorded by photographing a scintillator screen. Analysis of the obtained angular distributions was done using the FLUX computer simulation code [83] (see Subsect. 2.3.3) and the rainbow proton-atom interaction potential, introduced in Subsect. 4.1.3. The analysis included the calculations of the spatial distributions of transmitted protons for the chosen values of $E_0$. It was found that the width of the spatial distribution was minimal for $E_0 = 0.860$ MeV, which corresponded to the superfocusing point in the first rainbow cycle, i.e., $\Lambda = 0.250$. The resulting values of $\Lambda$ corresponding to the chosen values of $E_0$ were 0.232, 0.244, 0.251, 0.268, 0.277, and 0.288, respectively. The resulting average FWHM of the spatial distribution, i.e., the average width of the channeled proton beam at the exit from the crystal, at the superfocusing point, was ~20 pm. Therefore, since one of the measurements, for $\Lambda = 0.251$, was performed very close to the superfocusing point, one can state that the obtained experimental angular distribution together with the theoretical spatial distribution generated for $\Lambda = 0.250$ represent an indirect evidence of existence of the superfocusing effect.

## 4.4 Doughnut Effect

### *4.4.1 Explanation of the Doughnut Effect*

Let us now explore the crystal rainbow effect with protons of incident kinetic energy of $E_0 = 10$MeV impinging on a tilted <111> Si crystal [163–165]. The reduced crystal thickness has been chosen to be $\Lambda = 0.15$, meaning that the crystal is very thin. The angular distributions of transmitted protons and the corresponding rainbow patterns are obtained applying the theory of crystal rainbows in the same way as in the case of 60 MeV $Ne^{10+}$ ions and <111> Si crystal described in Sect. 3.4. As a result, the value of the crystal thickness in question is $L = 121$ nm. The crystal tilt angle, $\varphi$, is varied from zero up to the critical angle for axial channeling, $\psi_c = 2.62$ mrad, determined by Eq. (2.12). The crystal is tilted about the $x$ axis. The sizes of a bin along the $\Theta_x$ and $\Theta_y$ axes both equal 0.048 mrad. The incident number of protons is 258,788.

Figure 4.21 shows the angular distributions of 10 MeV protons transmitted through the <111> Si crystal obtained for $\varphi = 0$, $0.2\psi_c$, $0.4\psi_c$, and $0.6\psi_c$ [163, 164]. The distribution for $\varphi = 0$ contains a cusped hexagonal part in the central region of the TA plane with the cusps lying along the lines $\Phi = \tan^{-1}(\Theta_y/\Theta_x) = n\pi/3$, $n = 0-5$. The distribution also contains six parabolic parts in the peripheral region of the TA plane lying along the lines $\Phi = (2n + 1)\pi/6$, $n = 0-5$. For $\varphi = 0.2\psi_c$, the distribution is characterized by a distorted cusped hexagonal part in the central region of the TA plane connected to a distorted parabolic part lying along the line $\Phi = \pi/2$. The distribution for $\varphi = 0.4\psi_c$ contains a circular part of a smaller radius connected at the origin to a distorted parabolic part lying along the

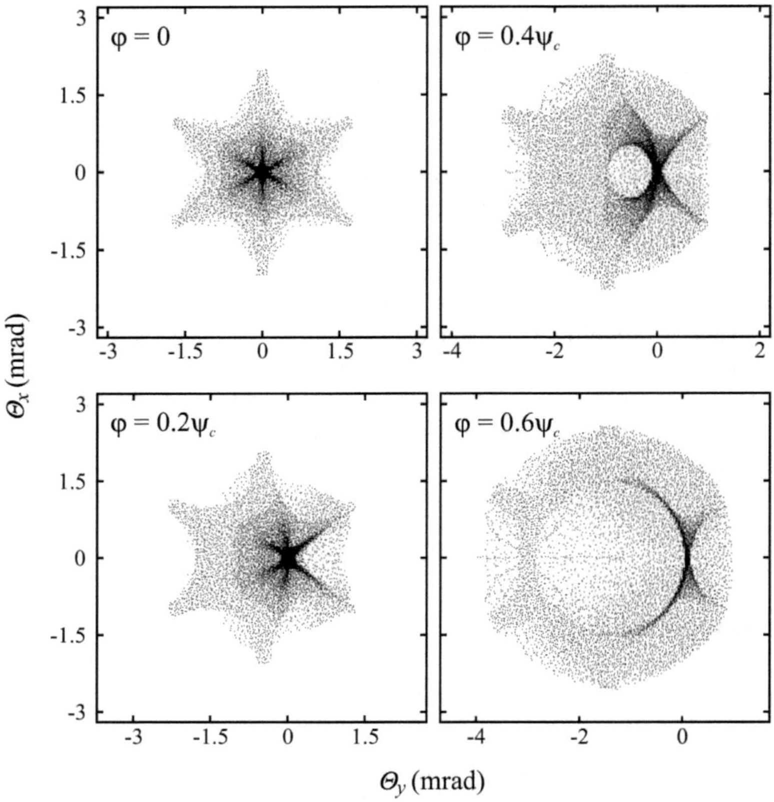


**Fig. 4.21** Angular distributions of 10 MeV protons transmitted through a tilted 121-nm thick <111> Si crystal for $\varphi = 0$, $0.2\psi_c$, $0.4\psi_c$, and $0.6\psi_c$[164]

line $\Phi = \pi/2$ and to a semicircular part of a larger radius lying along the line $\Phi = 3\pi/2$. For $\varphi = 0.6\psi_c$, the distribution is characterized by a circular part of a larger radius connected at the origin to a parabolic part lying along the line $\Phi = \pi/2$. The yield of transmitted protons along the circular part is not uniform. Its absolute maximum is at the origin, i.e., in the direction of the incident proton velocity vector, and its absolute minimum is on the line $\Phi = 3\pi/2$. For $\varphi > \psi_c$, the circular and distorted parabolic parts of the angular distribution disappear, leaving only a maximum in the direction of the incident proton velocity vector. This effect has been observed by a number of authors, e.g., by Chadderton [166], Armstrong et al. [112, 113], Rosner et al. [167], and Andersen et al. [168]. It has been referred to as the doughnut effect in ion channeling. However, its origin has not been determined prior to the above-presented study.

The rainbow patterns in the TA plane corresponding to the angular distributions shown in Fig. 4.21 are given in Fig. 4.22. For $\varphi = 0$, there are two cusped equilateral triangular rainbow lines in the central region of the TA plane with the cusps lying along the lines $\Phi = 2n\pi/3$ and $\Phi = (2n + 1)\pi/3$, $n = 0-2$, and six parabolic rainbow

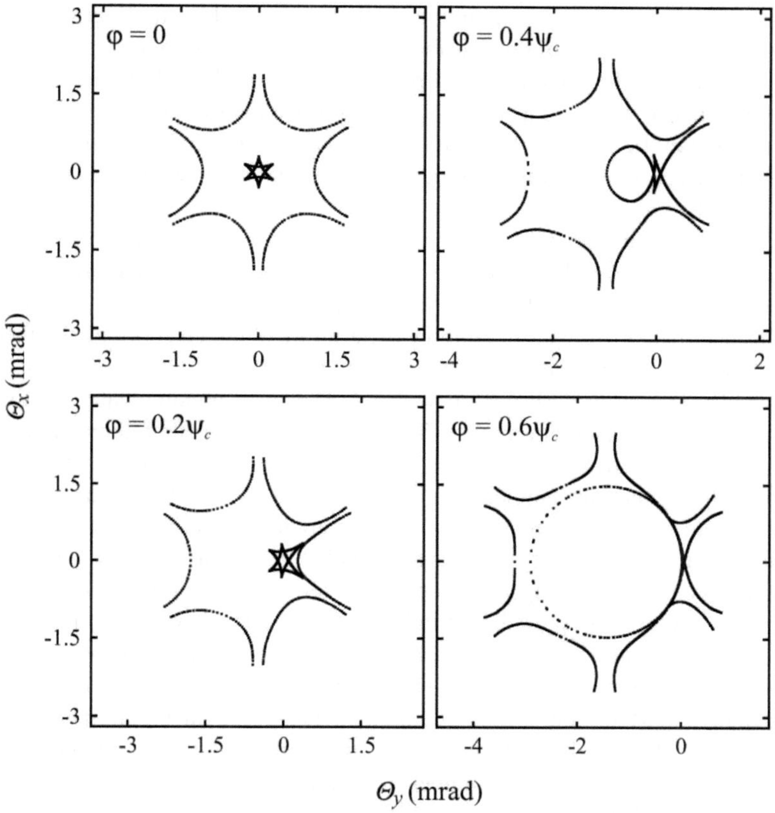


**Fig. 4.22** Rainbow lines in the TA plane for 10 MeV protons transmitted through a tilted 121-nm thick <111> Si crystal for $\varphi = 0$, $0.2\psi_c$, $0.4\psi_c$, and $0.6\psi_c$[164]

lines in the peripheral region of the TA plane lying along the lines $\Phi = (2n+1)\pi/6$, $n = 0-5$. The two triangular lines coincide with the lines shown in Fig. 3.32 for $\Lambda = 0.15$ while the six parabolic lines correspond to the cusped equilateral hexagonal line given in the same figure for $\Lambda = 0.20$. In the case under consideration, for $\Lambda = 0.15$, the hexagonal line has not yet been fully developed. The rainbow pattern for $\varphi = 0.2\psi_c$ clearly shows that the two cusped triangular lines and the parabolic line lying along the line $\Phi = \pi/2$ "attract" each other. For $\varphi = 0.4\psi_c$, this "interaction" results in a circular rainbow line of a smaller radius connected at the origin to a swallow-tailed parabolic rainbow line lying along the line $\Phi = \pi/2$ and being close to two distorted parabolic rainbow lines lying along the lines $\Phi = (2n+1)\pi/6$, $n = 0$ and 2. The rainbow pattern for $\varphi = 0.6\psi_c$ contains a circular line of a larger radius connected to three distorted parabolic lines lying along the lines $\Phi = (2n+1)\pi/6$, $n = 0-2$, and being close to three distorted parabolic lines lying along the lines $\Phi = (2n+1)\pi/6$, $n = 3-5$.

Comparison of Figs. 4.21 and 4.22 shows that the evolution of the angular distribution of transmitted protons with an increase of $\varphi$ can be fully explained by the evolution of the corresponding rainbow pattern. For the largest value of $\varphi$ ($0.6\psi_c$), the circular part of the distribution can be attributed to the circular rainbow line and partly to the three distorted parabolic rainbow lines lying along the lines $\Phi = (2n + 1)\pi/6$, $n = 0-2$. The parabolic part of the distribution coincides with the distorted parabolic rainbow line lying along the line $\Phi = \pi/2$. This means that the doughnut effect in ion channeling, occurring for the larger values of $\varphi$, is in fact a crystal rainbow effect.

We shall present here two additional cases of ion transmission through channels of a tilted crystal[169]. In the first case, the projectiles are 10 MeV protons and the target a 100-nm thick <110> Si crystal. The angular distributions of transmitted protons and the corresponding rainbow patterns are obtained applying the theory of crystal rainbows in the same way as in the cases of 60 MeV $Ne^{10+}$ ions and <100> and <111> Si crystals described in Sect. 3.4. The corresponding value of $\Lambda$ is 0.11, meaning that the crystal is very thin. This crystal is of the centered rectangular type with two atomic strings of the crystal per channel. The channel comprises two isosceles triangular subchannels. It is assumed that, when the crystal is not tilted, the subchannel axes intersect the $x$ axis. The number of atomic strings is 32, i.e., the atomic strings lying on the two nearest rhombic coordination lines are taken into account. The crystal tilt angle is varied from zero up to the critical angle for axial channeling, being now $\psi_c = 3.24$ mrad [Eq. (2.12)]. The crystal is tilted about the $x$ axis. The calculations are performed only for the subchannel whose axis lies on the positive part of the $x$ axis. The contribution of the subchannel whose axis lies on the negative part of the $x$ axis is obtained using the fact that the channel is symmetric relative to the $y$ axis. The sizes of a bin along the $\Theta_x$ and $\Theta_y$ axes both equal 0.08 mrad. The incident number of protons is 50,565.

The angular distribution of transmitted protons obtained for $\varphi = 0$ contains a characteristic hexagonal part lying along the $\Theta_y$ axis, whose shape resembles the shape of the distribution shown in Fig. 3.6 rotated by 90° about the origin because of the different orientation of the arrangement of atomic strings. The distribution is fully explained by the corresponding rainbow pattern in the TA plane, which contains a characteristic hexagonal line, whose shape is similar to the shape of the line shown in Fig. 3.5, four parabolic lines connected to the hexagonal line, and four parabolic lines away from it. For $\varphi = 0.95\psi_c$, the distribution consists of a circular part of a larger radius and a circular part of a smaller radius that are close to each other in the central region of the TA plane. The radius of the larger circular part is close to $0.95\psi_c$. The distribution also contains a pronounced maximum at the origin, i.e., in the direction of the incident proton velocity vector. Again, the distribution is fully explained by the corresponding rainbow pattern in the TA plane, which consists of a circular part of a larger radius defined by several shorter lines and a circular part of a smaller radius defined by two circular lines.

In the second additional case of ion transmission through channels of a tilted crystal to be presented here, the projectiles are 3.2 MeV protons and the target is a 190-nm thick <110> Si crystal. The crystal is tilted about the $x$ axis with $\varphi = 0.99\psi_c = 0.327°$ [Eq. (2.12)]. The angular distribution of transmitted protons and the corresponding rainbow pattern are obtained applying the theory of crystal rainbows in the same way as in the first additional case. The corresponding value of $\Lambda$ is 0.37, showing that the crystal is thin. These values of $E_0$, $L$, and $\varphi$ were chosen because the above-mentioned experimental study of Rosner et al. [167] included an excellent measurement of the doughnut effect with these parameters. In order to reproduce that measurement, we also generated the corresponding transmission pattern. It was obtained in a similar way as the angular distribution, but including the incident proton beam size and divergence angle, the proton collisions with the crystal's electrons, and the position of the proton detector as well. The proton collisions with the crystal's electrons are calculated by Eqs. (4.3) and (4.4). The beam diameter in the entrance plane of the crystal was 1 mm, its divergence angle (FWHM) 0.02°, and the distance of the detector from the crystal 84.8 cm. The transverse components of the initial proton position vectors within the beam and within the channel were chosen randomly using the Gaussian distribution and the uniform distribution, respectively. The sizes of a bin in the detector plane along the $x$ and $y$ axes were both 0.1 mm. The incident number of protons was 180,382.

The obtained theoretical and experimental transmission patterns are depicted in Figs. 4.23 and 4.24, respectively. Both transmission patterns contain a circular part of a radius close to $0.99\psi_c$ and a pronounced maximum at the origin, and, in both

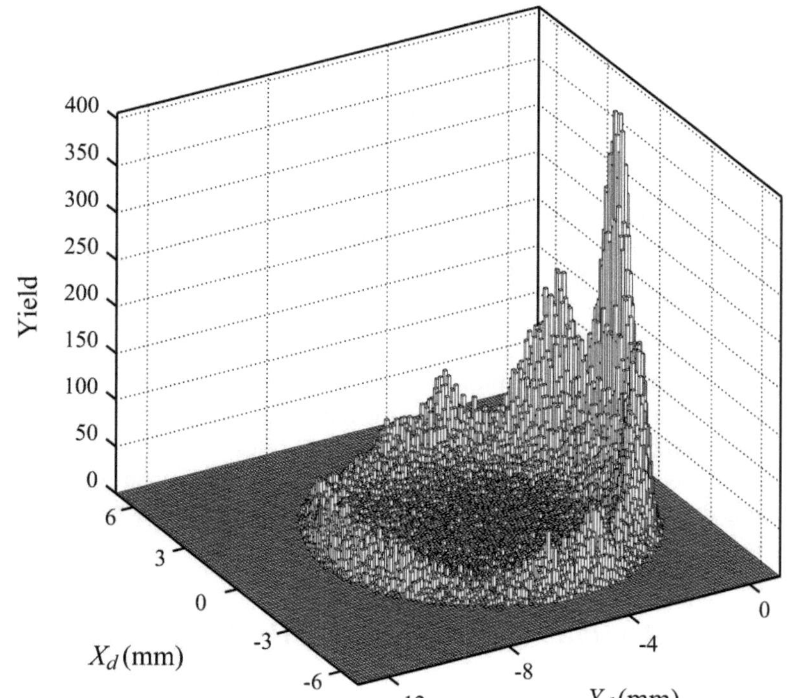


**Fig. 4.23** Theoretical transmission pattern of 3.2 MeV protons channeled in a tilted 190-nm thick <110> Si crystal for $\varphi = 0.99\psi_c$[169]. Variables $X_d$ and $Y_d$ are the vertical and horizontal components of the transverse position vector in the detector plane, respectively

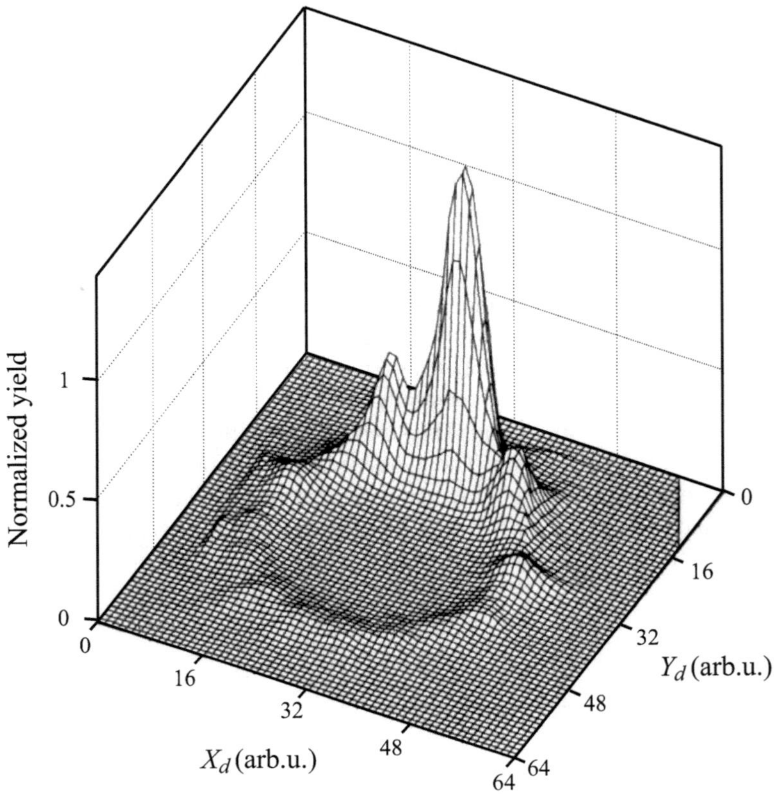


**Fig. 4.24** Experimental transmission pattern of 3.2 MeV protons channeled in a tilted 190-nm thick <110> Si crystal for $\varphi = 0.99\psi_c$[169]. Variables $X_d$ and $Y_d$ are the vertical and horizontal components of the transverse position vector in the detector plane, respectively; their values are given in arbitrary units

cases, the yield of transmitted protons along the circular part is not uniform. One can conclude that the agreement between the theoretical and experimental results is very good. It should be noted that the experimental study was conducted before the discovery of the crystal rainbow effect [40, 41], and that, consequently, the obtained results could not have been connected in any way to that effect.

### 4.4.2 High-Resolution Measurement of the Doughnut Effect

It has been shown in the previous subsection that the doughnut effect is in fact the rainbow effect with a tilted crystal. But how does the detailed evolution of a crystal rainbow into a doughnut look like? This subsection contains an answer to this question. It has been obtained on the basis of a sequence of high-resolution measurements with a 2.0 MeV focused proton microbeam transmitted through a 55-nm thick <100> Si crystal with $\varphi$ varying between 0 and 0.15° performed by the Singapore group and analyzed by us [130, 170]. The maximal value of $\varphi$ is equal to $0.43\psi_c$[Eq. (2.12)].

Figure 4.25 gives a sequence of experimental angular distributions of transmitted protons obtained for $\varphi = 0, 0.05, 0.06, 0.07, 0.09$, and 0.15°. The distribution for $\varphi = 0$ coincides with the first distribution shown in Fig. 4.4b. As $\varphi$ increases, the distribution changes gradually from a square-like to a ring-like distribution, emerging for $\varphi = 0.09°$ and being finally formed for $\varphi = 0.15°$. For the final value of $\varphi$, the yield of transmitted protons is maximal at the origin and decreases along the ring. The ring lies on the $\Theta_y$ axis and has four more and four less pronounced cusps on its outer edge. As it has been shown in the previous subsection, this is a doughnut. It should be noted that neither the square-like distributions preceding the ring-like ones nor the cusps on their outer edges were observed with the thicker crystals [112, 113, 166–168]. The figure also gives the corresponding rainbow patterns, which have been obtained applying the second part of the theory of crystal rainbows, presented in Sect. 3.4, with the continuum approximation [66]. They were first calculated with the ZBL proton–atom interaction potential [130]. However, the results shown in the figure have been obtained with the rainbow interaction potential [114], presented in Subsect. 4.1.3, which has proven to be more accurate than the ZBL potential in the case under consideration. The value of $\Lambda$ in question is 0.12, i.e., the crystal is very thin. When the crystal is not tilted, the atomic strings defining the channel intersect the lines $y = \pm x$. The number of atomic strings is 36, i.e., we take into account the atomic strings lying on the four nearest square coordination lines. The crystal is tilted about the $-x$ axis. The thermal vibrations of the crystal's atoms are included. One can clearly see that the rainbow patterns appear as the "skeletons" of the distributions. The central rainbow line, being a cusped square for $\varphi = 0$, changes to the cusped quasi-rectangles for $\varphi = 0.05$ and 0.06°, to the cusped transition forms for $\varphi = 0.07$ and 0.09°, and to a circle for $\varphi = 0.15°$. On the other hand, the peripheral rainbow pattern does not change much with $\varphi$. For $\varphi = 0.15°$, it contains three pairs of equivalent lines and two additional lines that almost join to make eight cusps. Analysis has shown that the inner side of the central rainbow line for $\varphi = 0$, 0.05, 0.06, or 0.07° is the bright side of the rainbow and its outer side the dark side of the rainbow, that for $\varphi = 0.15°$, the situation is opposite, and that for $\varphi = 0.09$, the inner side of this line is partly the bright side and partly the dark side of the rainbow and its outer side partly the dark side and partly the bright side of the rainbow. The inner side of the peripheral part of the rainbow pattern is the bright side of the rainbow and its outer side the dark side of the rainbow for all the values of $\varphi$.

We are now going to analyze the rainbow patterns in the IP plane corresponding to the rainbow patterns shown in Fig. 4.25. Figure 4.26a gives the rainbow pattern in the IP plane for $\varphi = 0$. It contains a line around the channel axis, designated as $a$, whose image in the TA plane is line $a'$ in Fig. 4.25a. The crossings of this line with the $y$ axis are designated as 1 and 2, and their images as $1'$ and $2'$, respectively. This rainbow pattern also contains eight lines around each atomic string defining the

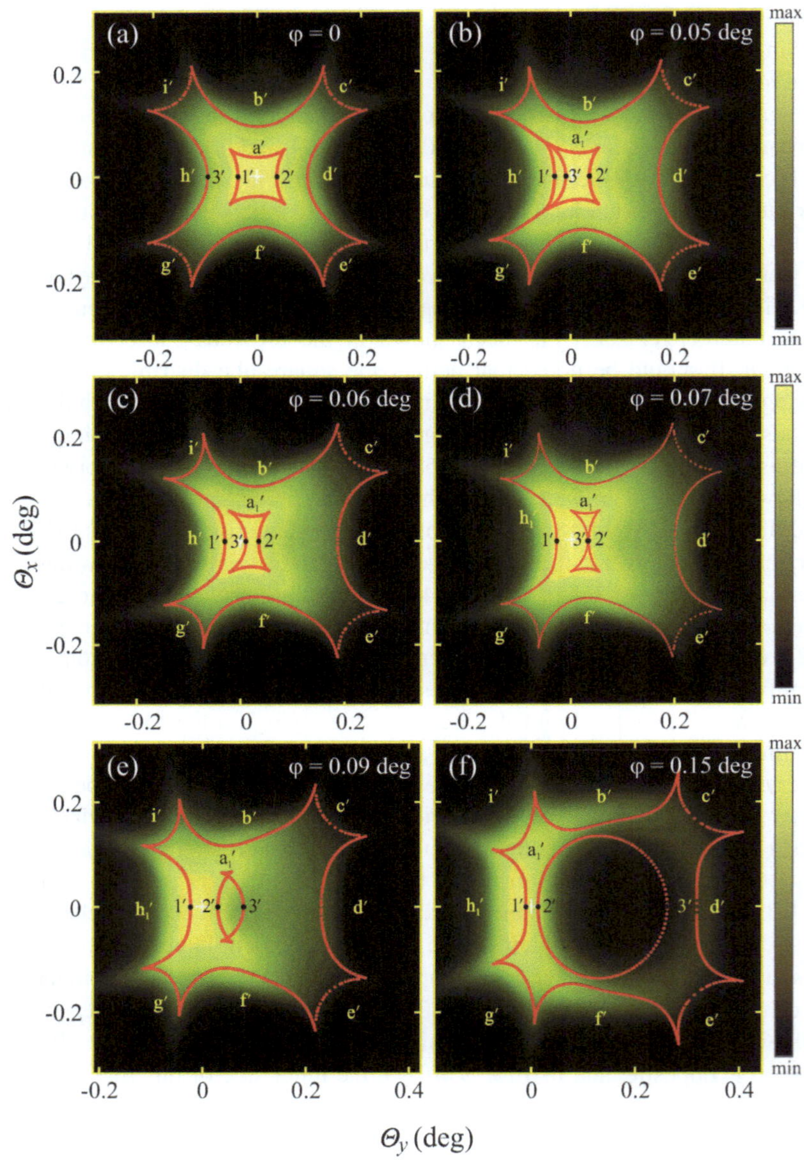


**Fig. 4.25** Experimental angular distributions of 2 MeV protons transmitted through a 55-nm thick <100> Si crystal for (**a**) $\varphi = 0°$, (**b**) $\varphi = 0.05°$, (**c**) $\varphi = 0.06°$, (**d**) $\varphi = 0.07°$, (**e**) $\varphi = 0.09°$, and (**f**) $\varphi = 0.15°$. The *red lines* are the associated rainbow lines in the TA plane (Adapted from Ref. [130])

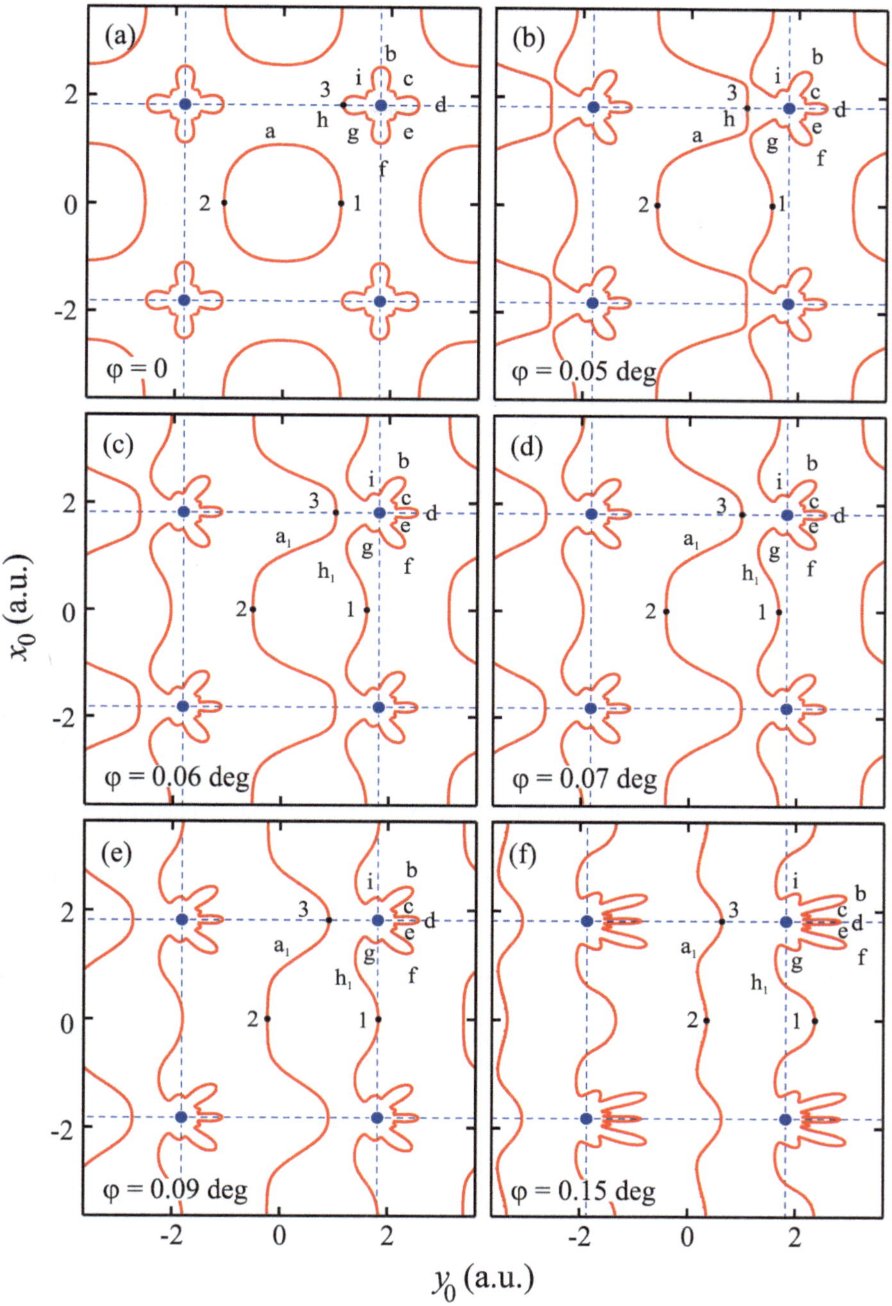


**Fig. 4.26** Rainbow lines in the IP plane for 2 MeV protons transmitted through a 55-nm thick <100> Si crystal for (**a**) $\varphi = 0°$, (**b**) $\varphi = 0.05°$, (**c**) $\varphi = 0.06°$, (**d**) $\varphi = 0.07°$, (**e**) $\varphi = 0.09°$, and (**f**) $\varphi = 0.15°$. The *blue full circles* represent the atomic strings defining the channel (Adapted from Ref. [130])

channel, designated as *b*-*i*, whose images in the TA plane are lines $b'$-$i'$ in Fig. 4.25a, respectively. The crossing of line $h$ with the line parallel to the $y$ axis connecting the two upper atomic strings defining the channel is designated as 3 and its image as 3′. A detailed analysis of the evolution of this rainbow pattern in the IP plane has shown that, as $\varphi$ increases, the "interaction" of the lines it contains with each other becomes evident. In particular, lines $a$ and $h$ "attract" each other. For $\varphi$ between 0 and 0.05°, these lines join dividing line $a$ into its left and right parts and line $h$ into its left, upper, and lower parts. As a result, instead of line $a$, one obtains line $a_1$, being composed of the left part of line $a$ and left part of line $h$, and line $h_1$, being composed of the right part of line $a$ and upper and lower parts of line $h$. This is seen in Fig. 4.26b, which gives the rainbow pattern in the IP plane for $\varphi = 0.05°$. It is evident that line $a_1$connects the neighboring channels and line $h_1$the neighboring atomic strings along the $x$ axis. The rainbow patterns in the IP plane for $\varphi = 0.06$, 0.07, 0.09, and 0.15° are depicted in Figs. 4.26c–f, respectively. For $\varphi = 0.15°$, line $a_1$ is close and almost parallel to the $x$ axis.

We are now returning to the rainbow patterns shown in Fig. 4.25, which are the images of the rainbow patterns given in Fig. 4.26. For $\varphi = 0$, line $a'$ contains points 1′ and 2′, and line $h'$ point 3′. However, for $\varphi = 0.05°$, i.e., after the joining of lines $a'$ and $h'$, point 1′ belongs to line ${h_1}'$ and point 3′ to line${a_1}'$, with point 3′ being closer to the origin than point 2′. The region between these two points, being on the right side of point 3′ and the left side of point 2′, is the bright side of the rainbow. As $\varphi$ increases further to 0.06, 0.07, 0.09, and 0.015, points 2′ and 3′ move toward each other, join, and move away from each other. For $\varphi = 0.09$ and 0.15°, point 3′ is farther from the origin than point 2′. Now, the region between these two points, being on the right side of point 2′ and the left side of point 3′, is the dark side of the rainbow. This region is an analogue of Alexander's dark band, occurring between the primary and secondary meteorological rainbows [2]. This has been the above-anticipated answer to the question about the detailed evolution of a crystal rainbow to a doughnut. We consider its contents fascinating.

Let us also mention that, in accordance with the discussion in the final part of Subsect. 4.1.3, the observed central and peripheral rainbows in the TA plane shown in Fig. 4.25a can be classified in the standard way as the primary and secondary rainbows, respectively. However, the observed central and peripheral rainbows in this plane presented in Figs. 4.25b–f cannot be classified in the standard way.

# Chapter 5
# Rainbows with Protons and Carbon Nanotubes

Carbon nanotubes were discovered by Iijima in 1991 [171]. They can be described as sheets of carbon atoms rolled up into cylinders with the atoms lying on the hexagonal lattice sites. The diameters of nanotubes are of the order of a few nanometers, and their lengths can be above a hundred micrometers. Nanotubes can be single-walled and multiwalled ones, depending on the number of cylinders they include. A nanotube is achiral or chiral. If it is achiral, the nanotube includes the atomic strings parallel to its axis. If it is chiral, the nanotube contains the atomic strings that spiral around its axis. Nanotubes have remarkable geometrical and physical properties [172]. As a result, they have begun to play an important role in the field of nanotechnologies [173].

Soon after the discovery of carbon nanotubes, Klimov and Letokhov [174] predicted that they could be used to channel positively charged particles. After that, a number of theoretical groups have studied ion channeling in nanotubes [49, 132, 175–206]. The main objective of most of those studies was to investigate the possibilities of guiding ion beams with nanotubes. Biryukov and Bellucci [177] looked for the nanotube diameter optimal for channeling high-energy ion beams. Krasheninnikov and Nordlund [183] studied the channeling of low-energy $Ar^+$ ions in achiral and chiral nanotubes. Petrović et al. [49] demonstrated that the rainbow effect could play an important role in ion transmission through nanotubes. Experimental studying of ion channeling in nanotubes is in an initial phase. The most challenging task in such experiments still is to solve the problems of ordering, straightening, and holding nanotubes. The first experimental data on ion channeling in nanotubes were reported by Zhu et al. [207]. They were obtained with $He^+$ ions of incident kinetic energy of 2 MeV and an array of well-ordered multiwall carbon nanotubes grown in a porous anodic aluminum oxide ($Al_2O_3$) membrane. The authors measured and compared the yields of ions transmitted through a bare $Al_2O_3$ sample and an $Al_2O_3$ sample including nanotubes. The first experiment with electrons and nanotubes was performed by Chai et al. [208]. They used 300 keV electrons and studied their transport through multiwall carbon nanotubes



N. Nešković et al., *Rainbows in Channeling of Charged Particles in Crystals and Nanotubes*, Lecture Notes in Nanoscale Science and Technology 25,
DOI 10.1007/978-3-319-61524-0_5

of lengths in the range of 0.7–3.0 μm embedded in a carbon fiber coating. Berdinsky et al. [209] succeeded in growing single-wall carbon nanotubes in etched ion tracks in a silicon oxide ($SiO_2$) layer on a silicon substrate. This result has opened a possibility for conducting precise measurements of ion channeling in nanotubes in a wide range of its kinetic energy.

In the period between 2003 and 2010, our group was intensely investigating the process of proton transmission through carbon nanotubes. The most important of the produced results were included in the booklet entitled *Channeling of Protons through Carbon Nanotubes* [210]. We analyzed the angular distributions and the corresponding rainbow patterns of 14–176 MeV and 1 GeV protons transmitted through straight very short, short, and long bundles of (10, 10) achiral single-wall nanotubes [49, 181, 186, 188, 190]. The chosen targets were also a bent short bundle of (10, 10) nanotubes [185] and straight very short and short bundles of (10, 0) achiral single-wall nanotubes [187] all being exposed to 1 GeV protons. The former study [185] included the effect of zero-degree focusing of 1 GeV protons channeled in a straight bundle of (10, 10) nanotubes. Then, we investigated the angular distributions generated with 1 GeV protons and straight very long (10, 0) and (10, 10) nanotubes [198, 204], and the angular distributions and the corresponding rainbow patterns obtained with protons of the same kinetic energy and straight very long (11, 9) chiral single-wall nanotubes [201]. Further, we explored the angular and spatial distributions and the corresponding rainbow patterns in the case of 1 GeV protons and a bent very short (11, 9) nanotube [206]. The obtained results suggested that the rainbow effect, which was clearly seen in the angular and spatial distributions, could be used for characterization of bundles of nanotubes – for deducing their transverse geometrical structure and measuring the average electron density inside and in between nanotubes. In addition, it was found that the effect could be used for shaping ion beams bent with nanotubes to be employed within high-energy accelerator facilities.

Besides, our studies comprised the influence of the effect of dynamic polarization of the carbon atoms' valence electrons on the angular and spatial distributions of protons transmitted through (11, 9) carbon nanotubes and the corresponding angular and spatial rainbows [189, 195–197]. The magnitude of the incident proton velocity vector was varied between 3 and 10 a.u. and the nanotube length between 0.1 and 1.0 μm. The nanotubes were placed in vacuum or embedded in the dielectric media – $SiO_2$, $Al_2O_3$, and Ni. We also explored the cases in which protons were directed into (6, 4) in (11, 9) chiral double-wall nanotubes [191] and a tilted (11, 9) nanotube [205]. In the latter case, the doughnut effect in proton channeling was observed. Our conclusions were that the rainbow effect could be used for measuring the electron distribution within a nanotube, especially the influence of dielectric media on the electron distribution, being relevant for employing nanotubes in nanoelectronics, and for shaping nanosized ion beams created with nanotubes embedded in dielectric media to be used for research in materials science and biomedicine. An additional conclusion was that the effect could be used for locating atoms or molecules intercalated in nanotubes.

In this chapter, we shall describe the zero-degree focusing effect in proton channeling in straight bundles of achiral nanotubes and the rainbow effect in proton transmission through a bundle of straight very short achiral nanotubes. The results produced in the investigation of proton transmission through a bent very short chiral nanotube will be also presented. Finally, we shall analyze in detail the spatial and angular rainbow effects in proton transmission through straight very long chiral nanotubes.

## 5.1 Rainbows with a Straight Very Short Bundle of Nanotubes

### 5.1.1 Zero-Degree Focusing Effect

Let us first consider the effect of zero-degree focusing of protons of incident kinetic energy of $E_0 = 1$ GeV channeled in straight bundles of (10, 10) achiral single-wall carbon nanotubes [185, 186]. This effect has been described in Sect. 3.3. It is assumed that the transverse cross-section of the bundle can be descirbed via a (two-dimensional) hexagonal or rhombic superlattice with one nanotube per primitive cell [211]. We choose the $z$ axis of the reference frame to coincide with the bundle axis, passing through the center of the primitive cell. The incident proton velocity vectors are all taken to be parallel to the bundle axis. The arrangement of the nanotubes defining the primitive cell is such that their axes intersect the $x$ and $y$ axes of the reference frame, which are the vertical and horizontal axes, respectively, with the nanotubes defining the longer diagonal of the primitive cell intersecting the $x$ axis. The origin of the reference frame lies in the entrance plane of the bundle. This is shown in Fig. 5.2. We take into account the contributions of the nanotubes lying on the two nearest rhombic coordination lines, relative to the center of the primitive cell.

The calculations are performed using the first part of the theory of crystal rainbows (Sect. 3.4). The interaction of the proton and a nanotube atom is described by Molière's approximation of the Thomas-Fermi interaction potential determined by Eqs. (2.1), (2.2), and (2.4), with $Z_1 = 1$ and $Z_2 = 6$. It has been demonstrated that this expression provides excellent agreement with experimental results in the field of ion channeling [44]. We employ the continuum string model, described in Subsect. 2.3.2, which leads to the proton-bundle continuum interaction potential that reads

$$V(x, y) = \sum_{i=1}^{I} \sum_{j=1}^{J} V_{ij}(x, y), \tag{5.1}$$

where, in accordance with Eqs. (2.13), (2.14), (2.15), and (2.16),

$$V_{ij}(x,y) = \frac{2Z_1Z_2e^2}{d}\sum_{\ell=1}^{3}\alpha_\ell K_0\left(\frac{\beta_\ell \rho_{ij}}{a_{\mathrm{TF}}}\right) \tag{5.2}$$

is Molière's continuum interaction potential of the proton and the $j$th atomic string of the $i$th nanotube within the bundle, $d = 0.24$ nm is the distance between the atoms of a string [172], $\rho_{ij}^2 = \left(x - x_{ij}\right)^2 + \left(y - y_{ij}\right)^2$ is the distance between the proton and the string squared, $x$ and $y$ are the transverse components of the proton position vector, $x_{ij}$ and $y_{ij}$ are the transverse coordinates of the string, $I = 16$ is the number of nanotubes within the bundle, and $J = 40$ is the number of strings of a nanotube. Hence, the number of atomic strings within the bundle is $I \times J = 640$. The nanotube radius is $R_n = 0.67$ nm [172], and the distance between the axes of two neighboring nanotubes is 1.70 nm [211]. The thermal vibrations of the nanotube atoms are taken into account. In a way analogous to the one leading to Eq. (3.13), this is done by substituting $V_{ij}(x,y)$ in Eq. (5.1) with

$$V_{ij}^{\mathrm{th}}(x,y) = V_{ij}(x,y) + \frac{\sigma_{\mathrm{th}}^2}{2}\left[\partial_{xx}V_{ij}(x,y) + \partial_{yy}V_{ij}(x,y)\right], \tag{5.3}$$

where $\sigma_{\mathrm{th}} = 5.3$ pm is the one-dimensional atomic thermal vibration amplitude [212]. Thus, $V^{\mathrm{th}}(x,y)$ is obtained instead of $V(x,y)$. The electronic proton energy loss as well as the uncertainty of the proton channeling angle due to its collisions with the nanotube electrons are neglected. The components of the proton impact parameter vector are chosen uniformly from the uniform distribution within the primitive cell of the (rhombic) superlattice.

The nanotube walls define two separate regions in the TP plane: inside the nanotubes and in between them [49]. Accordingly, the bundle contains two types of channels: the circular one, whose center coincides with the center of the region inside each nanotube, and the triangular one, whose center coincides with the center of the region in between each three neighboring nanotubes. Therefore, in this case, one has to introduce two frequencies of proton oscillations around the channel axis, corresponding to the circular and triangular channels. Consequently, for each value of the bundle length, $L$, there are two values of the reduced bundle length, $\Lambda_1$ and $\Lambda_2$, corresponding to the proton motions close to the axes of the circular and triangular channels, respectively. The values of $\Lambda_1$ and $\Lambda_2$ are calculated by Eq. (3.16) in which $v_{\mathrm{av}}$ is taken to be the magnitude of the incident proton velocity vector, and $f_1 = f_{h1}$ and $f_2 = f_{h2}$ are deduced from the second order terms of the Taylor expansions of the proton-bundle continuum interaction potential in the vicinities of the circular and triangular channel centers, respectively.

Figure 5.1a shows the dependence of the zero-degree yield of protons propagating along the circular channels of the bundle on $L$ in the range of 0–200 μm; the $\Lambda_1$axis is shown too. For the region in the TA plane in the vicinity of the origin, we take the region in which the proton transmission angle, $\Theta = \left(\Theta_x^2 + \Theta_y^2\right)^{1/2}$, where $\Theta_x$ and $\Theta_y$ are its vertical and horizontal components, respectively, is smaller than

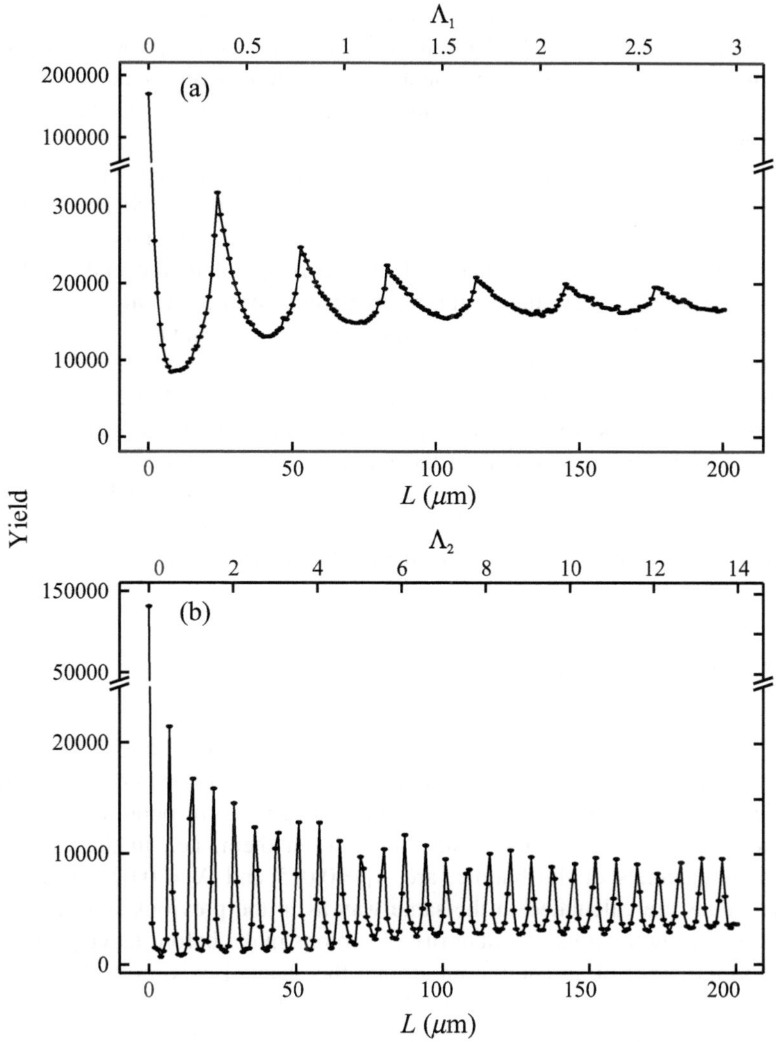


**Fig. 5.1** (**a**) Zero-degree yield of 1 GeV protons transmitted through the circular channels of a bundle of (10, 10) achiral single-wall carbon nanotubes as a function of $L$ [185]. (**b**) Corresponding zero-degree yield of protons transmitted through the triangular channels of the bundle as a function of $L$ [185]

0.0109 mrad. The incident number of protons is 174,976. The six maxima of the dependence correspond to the ends of the first six rainbow cycles, where the proton beam channeled in the circular channels is quasiparallel [91, 95]. The positions of these maxima should be close to $\Lambda_1 = 0.50, 1.00, 1.50, \ldots$. However, this is not true. For example, the position of the first maximum of the dependence is $\Lambda_1 = 0.35$. The large deviations of the values of these positions from their expected values are explained by a strong anharmonicity of the proton-bundle continuum interaction potential in the vicinities of the centers of the circular channels [95]. If the chosen region in the TA plane in the vicinity of the origin were smaller, the anharmonicity of the interaction potential would be less pronounced, and, consequently, the value of $\Lambda_1$ corresponding to the first maximum of the dependence would be closer to 0.50.

The dependence of the zero-degree yield of protons propagating along the triangular channels of the bundle on $L$ in the range of 0–200 μm is given in Fig. 5.1b; the $\Lambda_2$axis is also shown. The incident number of protons is 136,658. The dependence has 27 maxima, and they correspond to the ends of the first 27 rainbow cycles, where the proton beam channeled in the triangular channels is quasiparallel [91, 95]. The positions of these maxima are close to $\Lambda_2 = 0.50, 1.00, 1.50, \ldots$. For example, the position of the first maximum of the dependence is $\Lambda_2 = 0.49$. The small deviations of the values of these positions from their expected values are explained by a weak anharmonicity of the proton-bundle continuum interaction potential in the vicinities of the centers of the triangular channels [95]. It should be noted that the total zero-degree yield of channeled protons, i.e., of protons channeled in both the circular and triangular channels, as a function of $L$ is the sum of the zero-degree yields shown in Figs. 5.1a, b.

### 5.1.2 Rainbow Effect

Now, we shall describe the rainbow effect in the transmission of 1 GeV protons through a bundle of straight (10, 10) achiral single-wall carbon nanotubes [49]. The bundle length is $L =$ μm. The reduced bundle lengths corresponding to the proton channeling inside the nanotubes and in between them, i.e., in the circular and triangular channels of the bundle, are $\Lambda_1 = 0.015$ and $\Lambda_2 = 0.070$, respectively. These values are calculated in the same way as in the previous subsection. They tell us that, in both cases, the majority of protons make before leaving the bundle less than a quarter of an oscillation around the channel axis (both $\Lambda_1$ and $\Lambda_2$ are below 0.250). Therefore, one can say that the bundle we investigate is very short.

The calculations are performed using the theory of crystal rainbows (Sect. 3.4). The interaction of the proton and bundle is described by Eqs. (5.1) and (5.2). We take into account the thermal vibrations of the nanotube atoms, and this is done by Eq. (5.3). However, the proton collisions with the nanotube electrons are disregarded. This is justified by the fact that the bundle is very short. If the bundle were long (with $\Lambda_1$ and $\Lambda_2$ being above 1), this effect would cause a smearing out of the angular distributions of channeled protons [80, 204].

Figure 5.2 shows the rainbow lines in the IP plane. One can see that inside each nanotube, i.e., inside each circular channel of the bundle, there is one (closed) rainbow line, while in between each three neighboring nanotubes, i.e., inside each triangular channel of the bundle, there are one larger and four smaller (closed) rainbow lines.

The rainbow line in the TA plane that is the image of the rainbow line in the IP plane lying inside each nanotube is shown in Fig. 5.3a. It consists of 20 connected

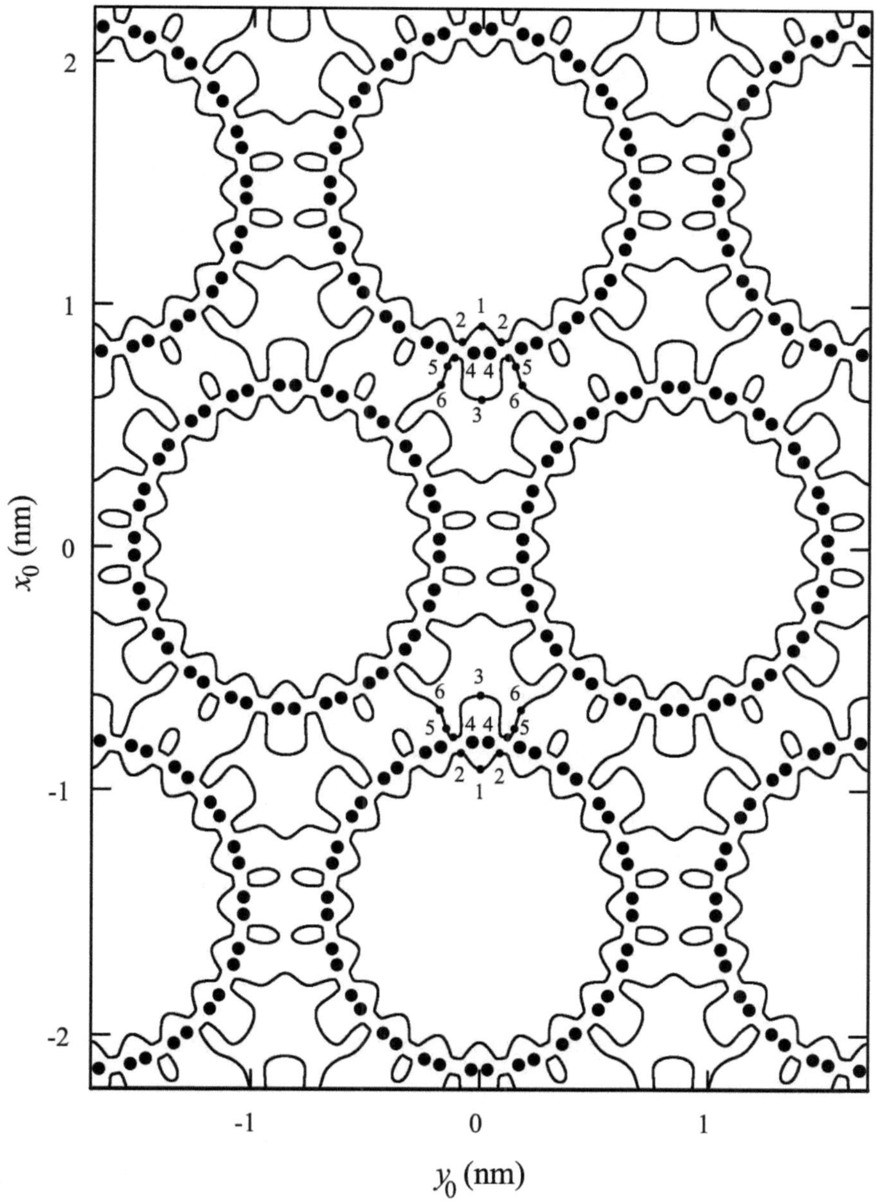


**Fig. 5.2** Rainbow lines in the IP plane for 1 GeV protons channeled in a 1-μm long bundle of (10, 10) achiral single-wall carbon nanotubes [49]

cusped triangular lines lying along the lines $\Phi = \tan^{-1}(\Theta_y/\Theta_x) = 2(n+1)\pi/20$, $n = 0-19$, which correspond to the parts of the rainbow line in the IP plane in front of the 20 pairs of atomic strings defining the nanotube (see Fig. 5.2). Points $1'$ and $2'$ are the intersection points of the rainbow line in the TA plane with the line $\Theta_y = 0$: points $1'$ are the apices of the cusps and points $2'$ are the intersections of the

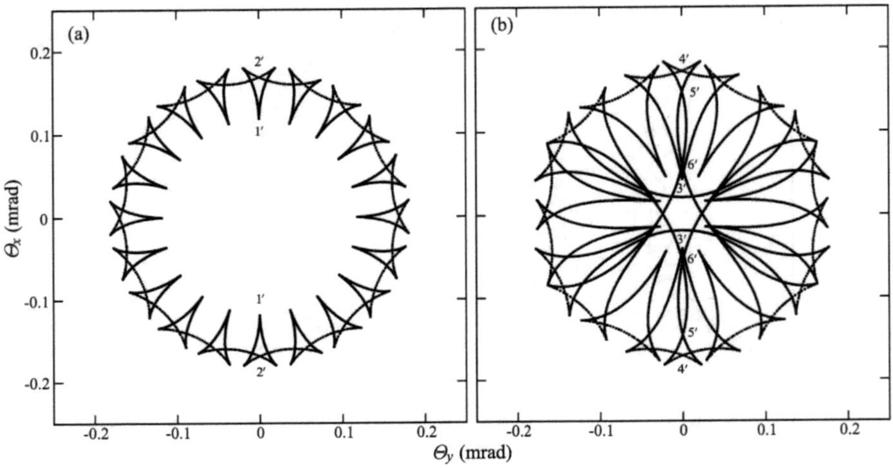


**Fig. 5.3** Rainbow lines in the TA plane corresponding to the rainbow lines in the IP plane shown in Fig. 5.2 lying (**a**) inside each nanotube and (**b**) in between each four neighboring nanotubes [49]

parts of the rainbow line with each other. The corresponding points in the IP plane are denoted by 1 and 2, respectively. Figure 5.3b shows the rainbow lines in the TA plane that are the images of the rainbow lines in the IP plane lying in between each four neighboring nanotubes. Analysis shows that this rainbow pattern consists of two cusped equilateral triangular rainbow lines in the central region of the TA plane with the cusps lying along the lines $\Phi = 2n\pi/3$ and $\Phi = (2n+1)\pi/3$, $n = 0-2$, each connected with three pairs of cusped triangular rainbow lines lying along the same lines, and eight cusped triangular rainbow lines lying in between the six pairs of triangular lines. The two equilateral triangular lines, each connected with the three pairs of triangular lines, are the images of the two larger rainbow lines, while the eight triangular lines are the images of the eight smaller rainbow lines in the IP plane (see Fig. 5.2). Points $3'$, $4'$, $5'$, and $6'$ are the intersection points of the rainbow lines in the TA plane with line $\Theta_y$=0: points $3'$ are the apices of the cusps, and points $4'$, $5'$, and $6'$ are the intersections of the parts of the larger rainbow lines with each other. The corresponding points in the IP plane are denoted by 3, 4, 5, and 6, respectively.

The angular distribution of transmitted protons is shown in Fig. 5.4. The incident number of protons is 2,142,538. The areas in which the yields of transmitted protons are larger than 0.13, 0.26, and 0.39%, 1.3, 2.6, and 3.9%, and 13, 26, and 39% of the maximal yield are designated by the increasing tones of gray color. At the very low level of the yield, corresponding to the boundary yields of 0.13, 0.26, and 0.39% of the maximal yield, there are 20 triangular forms in the peripheral region of the TA plane, with the maxima lying on the lines $\Phi = 2(n+1)\pi/20$, $n = 0-19$. Further, at the low level of the yield, corresponding to the boundary yields of 1.3, 2.6, and 3.9% of the maximal yield, there is a hexagonal structure in the central region of the TA plane, with the maxima lying on the lines $\Phi = n\pi/3$, $n = 0-5$. Finally, at the high level of the yield, corresponding to the boundary

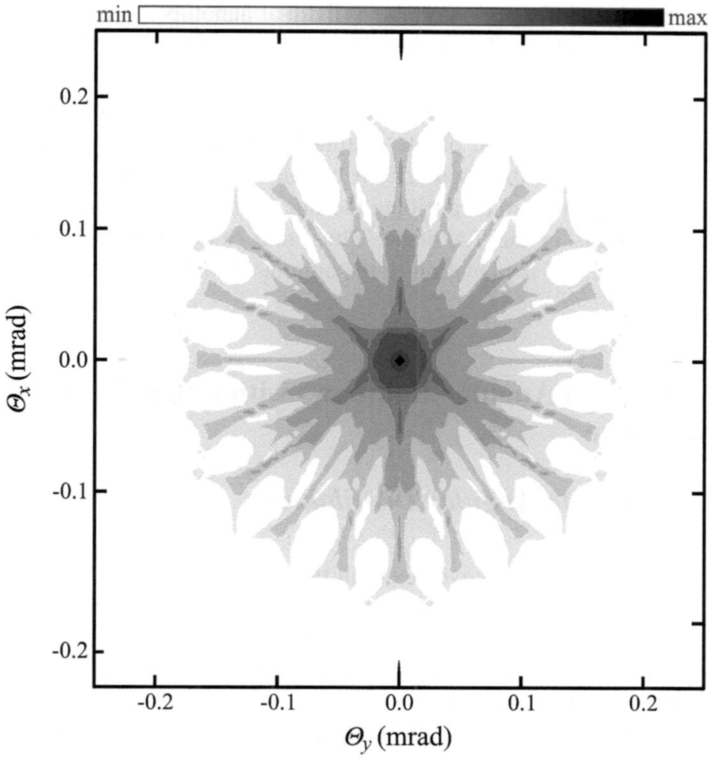


**Fig. 5.4** Angular distribution of 1 GeV protons transmitted through a 1-μm long bundle of (10, 10) achiral single-wall carbon nanotubes. The areas in which the yields of transmitted protons are larger than 0.13, 0.26, 0.39, 1.3, 2.6, 3.9, 13, 26, and 39% of the maximal yield are designated by the increasing tones of *gray color* (This is an adapted figure from Ref. [49])

yields of 13, 26, and 39% of the maximal yield, there is a pronounced maximum at the origin of the TA angle plane. The analysis shows that the first part of the angular distribution is generated by the protons with the impact parameters close to the atomic strings defining the nanotubes – the 20 triangular forms correspond to the 20 pairs of atomic strings defining the nanotubes. The second part of the angular distribution is generated by the protons with the impact parameters in between the nanotubes but not close to the centers of the triangular channels. The third part of the angular distribution is generated to a larger extent by the protons with the impact parameters close to the centers of the circular channels and to a smaller extent by the protons with the impact parameters close to the centers of the triangular channels. It should be noted that most of the protons that generate the third part of the angular distribution interact with the nanotubes very weakly – they move through the space inside the nanotubes virtually as through a drift space. Thus, we can say that the angular distribution contains the information on the transverse lattice structure of the bundle. Its first part (the peripheral region of the

TA plane) provides information on the individual nanotubes, while its second part (the central region of the TA plane) provides information on the way they are connected with each other.

Comparison of Figs. 5.3 and 5.4 clearly shows that the shape of the rainbow pattern determines the shape of the angular distribution of transmitted protons. Also, each maximum of the angular distribution, except the maximum lying at the origin of the TA plane, can be attributed to one of the above-mentioned characteristic rainbow points in the TA plane. Thus, one can conclude that the rainbow pattern enables the full explanation of the angular distribution.

## 5.2 Spatial and Angular Rainbows with a Bent Very Short Nanotube

### *5.2.1 Spatial and Angular Distributions*

The system we are going to investigate in this section is a proton moving along a bent (11, 9) chiral single-wall carbon nanotube [206]. We have chosen the incident proton kinetic energy to be $E_0 = 1$ GeV and the nanotube length $L = 7$µm. The nanotube contains 40 atomic strings, which spiral about its axis. The $z$ axis of the reference frame, being the longitudinal axis, is taken to coincide with the nanotube axis and its origin to lie in the entrance plane of the nantoube. The $x$ and $y$ axes of the reference frame, being the transverse axes, are the vertical and horizontal axes, respectively. This reference frame is an inertial reference frame. The incident proton velocity vectors are all taken to be parallel to the nanotube axis. The interaction of the proton and a nanotube atom is described by Molière's approximation of the Thomas-Fermi interaction potential given by Eqs. (2.1), (2.2), and (2.4). We employ the continuum approximation, described in Subsect. 2.3.2. However, since the nanotube under consideration is chiral, the proton-atom interaction potential is averaged both longitudinally and azimuthally. As a result, as it has been shown by Artru et al. [132], one obtains the proton-nanotube continuum interaction potential that reads

$$V_{\mathrm{ch}}(x, y) = V_{\mathrm{ch},0} \sum_{\ell=1}^{3} \alpha_\ell K_0\left(\frac{\beta_\ell R_n}{a_{\mathrm{TF}}}\right) I_0\left(\frac{\beta_\ell \rho}{a_{\mathrm{TF}}}\right), \tag{5.4}$$

with

$$V_{\mathrm{ch},0} = \frac{16\pi Z_1 Z_2 e^2 R_n}{3^{4/3}\, a_b^2}, \tag{5.5}$$

where $R_n = 0.69$ nm nm is the nanotube radius [172], $a_b = 0.14$ nm is the atom-atom bond length [172], $\rho = (x^2 + y^2)^{1/2}$, $x$ and $y$ are the transverse components of the proton position vector, and $I_0$ and $K_0$ denote the zero-order modified Bessel functions of the first and second kinds, respectively. It is evident that Eqs. (5.4) and (5.5) do not contain explicitly the nanotube chiral indices, $m = 11$ and $n = 9$. These indices are hidden in $R_n = [(3^{1/2}/(2\pi)](m^2 + mn + n^2)^{1/2}a_b$ [132]. With this interaction potential, the frequency of proton oscillations around the nanotube axis, $f = f_h$, is such that the reduced nanotube length is $\Lambda = 0.17$. This means that the nanotube is very short. It is evident that $V_{\rm ch}(x, y)$ is cylindrically symmetric. In order to simplify the treatment, we shall ignore the influence of the thermal vibrations of the nanotube atoms. Since the nanotube is very short, we shall also ignore the proton collisions with the nanotube electrons.

We shall apply the theory of crystal rainbows (Sect. 3.4). The components of the proton impact parameter vector are chosen randomly from the uniform distribution within a circular region around the origin of the IP plane of radius $R_n - a_{\rm TF}$. If the proton, during its motion along the nanotube, exits the circular region around the origin of the TP plane of radius $R_n - a_{\rm TF}$, it is excluded from the calculation. This is done because the proton motion becomes unstable when it approaches the nanotube wall, leading to its dechanneling, as it has been explained in Subsect. 2.3.2.

Let us now assume that the nanotube is bent along the $y$ axis. Its bending angle, $\alpha_b$, is varied between 0 and 2 mrad. In order to follow the proton motion properly, we introduce the reference frame that moves relative to the inertial reference frame along the bent nanotube axis with the velocity equal to the magnitude of the incident proton velocity vector, $v_0$. This means that the new reference frame rotates in the $yz$ plane along a circle of radius $R_b = L/\alpha_b$, being the radius of curvature of the nanotube, with the tangential velocity equal to $v_0$. This is a noninertial reference frame, in which one has to treat the nanotube as if it is straight and use the effective proton-nanotube continuum interaction potential,

$$V_{\rm ch}^{\rm eff}(x, y) = V_{\rm ch}(x, y) + \frac{mv_0^2 y}{R_b}, \quad (5.6)$$

instead of $V_{\rm ch}(x, y)$ [177]. The origin of the additional term in $V_{\rm ch}^{\rm eff}(x, y)$ is the inertial centrifugal force acting on the proton, occurring due to the bending of the nanotube. It makes $V_{\rm ch}^{\rm eff}(x, y)$ noncylindrically symmetric. As a result of introducing the noninertial reference frame, the final TP plane makes an angle with the IP plane equal to $\alpha_b$. We should introduce here Tsyganov's angle, which is the critical angle for bending of the nanotube. It is given by the relation $\psi_{\rm T} = L/R_c$, where $R_c = mv_0^2/(eE_c)$ is the critical radius of curvature of the nanotube, with $E_c$ being the electric field intensity at the minimal proton distance from the nanotube wall for which its motion is stable [180]. It has been established that the critical angle for channeling, introduced in Subsect. 2.3.2, satisfies the relation $\psi_c^2 = (2R_n/L)\psi_{\rm T}$ [180].

Figure 5.5 shows $V_{\rm ch}(x, y)$ and $V_{\rm ch}^{\rm eff}(x, y)$ for $\alpha_b = 0.2$ mrad along the $y$ axis. It is clear that along the positive part of this axis, i.e., toward the center of curvature of the nanotube, $V_{\rm ch}^{\rm eff}(x, y)$ is above $V_{\rm ch}(x, y)$, while along its negative part, i.e., away from the center of curvature, the situation is opposite. For the values of $y$ between $-0.61$ and 0 nm, $V_{\rm ch}^{\rm eff}(x, y)$ is negative, i.e., the proton is attracted to the outer

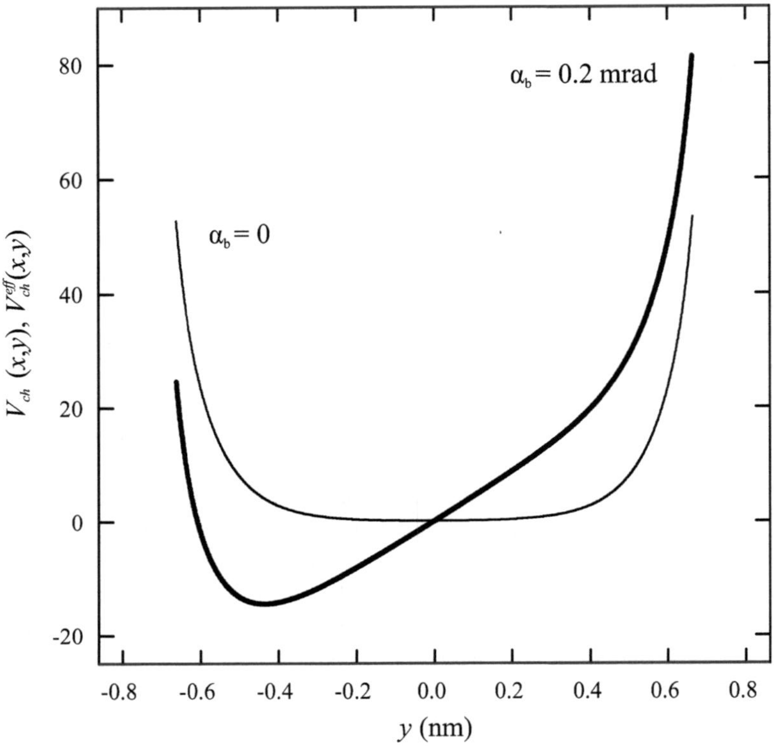


**Fig. 5.5** Continuum interaction potential along the $y$ axis of the proton and straight (11, 9) chiral single-wall carbon nanotube, $V_{\mathrm{ch}}(x,y)$ (*thin line*), and the corresponding effective continuum interaction potential of the proton and bent nanotube for $\alpha_b = 0.2$ mrad, $V_{\mathrm{ch}}^{\mathrm{eff}}(x,y)$ (*thick line*) [206]

nanotube wall, and for $y = -0.44$ nm, it is minimal. The difference between the two interaction potentials is smaller along the lines making the angles between 0 and 90° with the $y$ axis. When this angle equals 90°, $V_{\mathrm{ch}}^{\mathrm{eff}}(x,y)$ reduces to $V_{\mathrm{ch}}(x,y)$. All this is a consequence of the fact that the proton is exposed to the centrifugal force too, in addition to the atomic force.

The spatial distribution of protons transmitted through the straight nanotube, for $\alpha_b = 0$, is shown in Fig. 5.6. The incident number of protons is 785,662. Since $V_{\mathrm{ch}}(x,y)$ is cylindrically symmetric, the spatial distribution is rotationally symmetric. One can see that it contains a pronounced maximum at the origin and a circular ridge, whose radius is 0.17 nm. The transmitted proton beam is well confined within the circular ridge. As it must be, the limiting radius of the distribution is equal to $R_n - a_{\mathrm{TF}}$. The spatial distribution of protons transmitted through the bent nanotube for $\alpha_b = 0.2$ mrad is shown in Fig. 5.7. The incident number of protons is the same as in the case of $\alpha_b = 0$. Since $V_{\mathrm{ch}}^{\mathrm{eff}}(x,y)$ is not cylindrically symmetric, the spatial distribution is not rotationally symmetric. It contains a pronounced complex maximum within a quasi-ellipsoidal ridge close to the outer nanotube wall, which corresponds to the bottom of the interaction potential well defined by $V_{\mathrm{ch}}^{\mathrm{eff}}(x,y)$

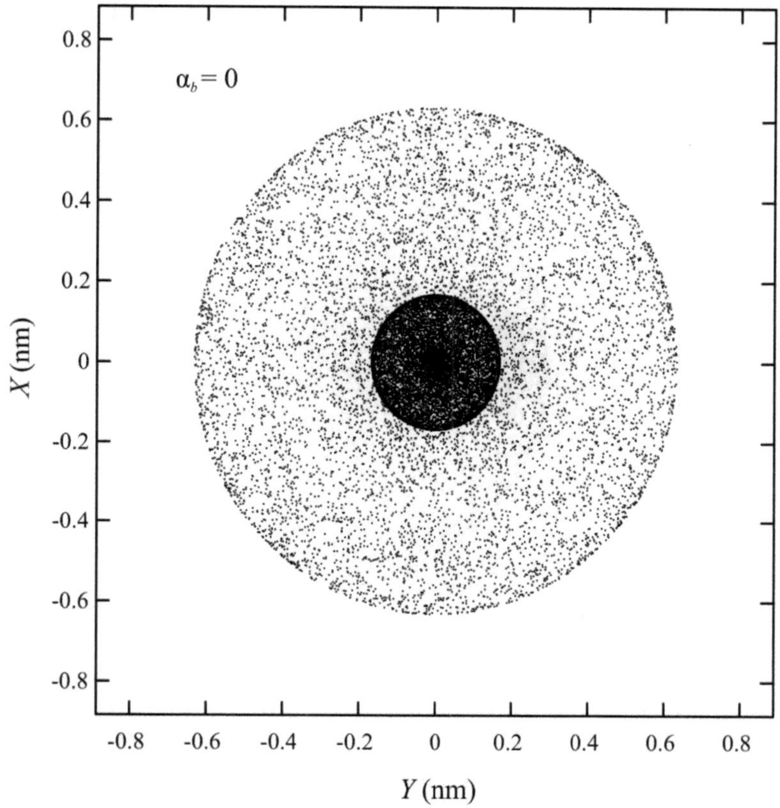


**Fig. 5.6** Spatial distribution of 1 GeV protons transmitted through a straight (11, 9) chiral single-wall carbon nanotube for $L = 7$ μm [206]

(see Fig. 5.5). The transmitted proton beam is well confined within the quasi-ellipsoidal ridge. Figure 5.8 shows the yields along the $Y$ axis of the protons transmitted through the straight nanotube and the bent nanotube for $\alpha_b = 0.2$ mrad. The former dependence demonstrates that the proton yield changes abruptly across the circular ridge appearing in the corresponding spatial distribution, with its absolute maximum lying at the origin. The latter dependence has two sharp maxima, which belong to the quasi-ellipsoidal ridge in the corresponding spatial distribution. The higher maximum is the absolute maximum of the spatial distribution.

The angular distribution of protons transmitted through the straight nanotube, for $\alpha_b = 0$, is given in Fig. 5.9. The incident number of protons is the same as in the cases of spatial distributions of transmitted protons. Since $V_{\text{ch}}(x, y)$ is cylindrically symmetric, the angular distribution is rotationally symmetric. It contains a pronounced maximum at the origin and a circular ridge, whose radius of 0.15 mrad. The transmitted proton beam is confined within the circular ridge. The limiting transmission angle of the distribution is 0.26 mrad. The angular distribution of

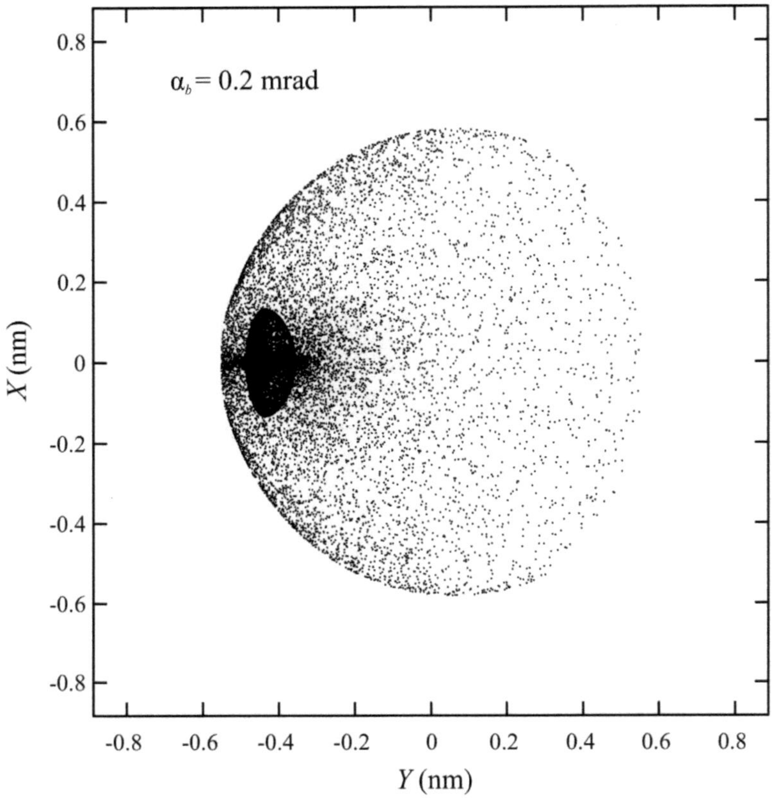


**Fig. 5.7** Spatial distribution of 1 GeV protons transmitted through a bent (11, 9) chiral single-wall carbon nanotube for $L = 7$ μm and $\alpha_b = 0.2$ mrad [206]

protons transmitted through the bent nanotube for $\alpha_b = 0.2$ mrad is given in Fig. 5.10. The incident number of protons is the same as in the case of $\alpha_b = 0$. Since $V_{\rm ch}^{\rm eff}(x, y)$ is not cylindrically symmetric, the angular distribution is not rotationally symmetric. It contains a maximum close to the outer nanotube wall and a half-moon-shaped ridge in the central region, all within an acorn-shaped ridge. This angular distribution is similar to the one obtained earlier with 1 GeV protons and a bundle of bent (10, 10) single-wall carbon nanotubes [185]. The transmitted proton beam is confined within the half-moon-shaped ridge and to a smaller extent within the rest of the acorn-shaped ridge. It is clear that the transmitted proton beam is split into two components. One can say that the component of the transmitted proton beam around the maximum close to the outer nanotube wall is its nonbent component and that its component within the half-moon-shaped ridge is its bent component. Let us not forget that the final TP plane makes the angle $\alpha_b$ with the IP plane. Figure 5.11 gives the yields along the $\Theta_y$ axis of the protons transmitted through the straight nanotube and the bent nanotube for $\alpha_b = 0.2$ mrad. The former dependence demonstrates that the proton yield changes abruptly across

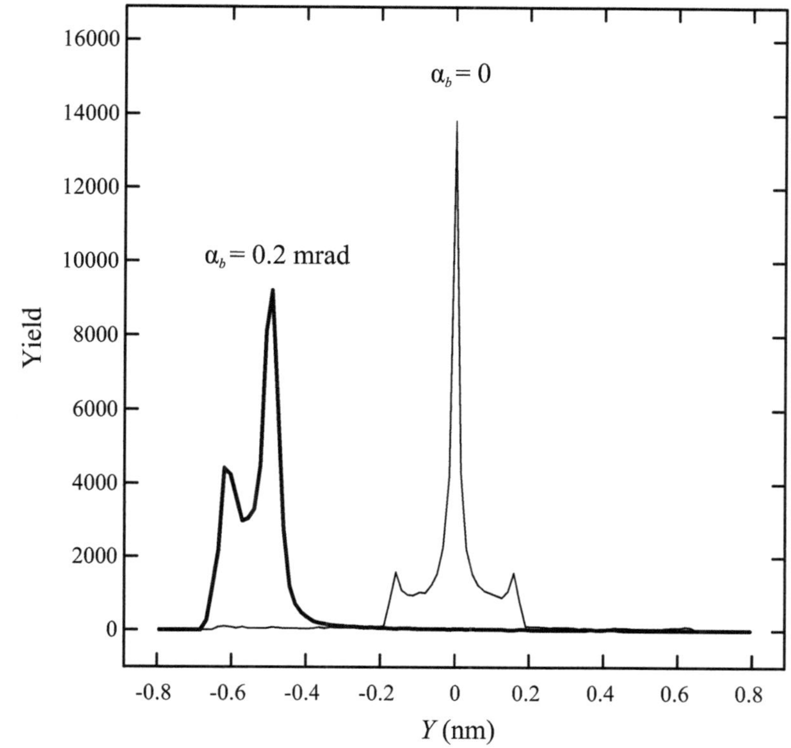


**Fig. 5.8** Yield along the $Y$ axis of 1 GeV protons transmitted through a straight (11, 9) chiral single-wall carbon nanotube for $L = 7$ μm (*thin line*) and the corresponding yield of protons transmitted through the bent nanotube for $\alpha_b = 0.2$ mrad (*thick line*) [206]

the circular ridge appearing in the corresponding angular dependence. The latter dependence has three sharp maxima. The highest maximum, lying around point $\Theta_y = -0.2$ mrad, represents the nonbent component of the transmitted proton beam and the pair of lower maxima, lying around point $\Theta_y = 0$, its bent component. The corresponding angular distribution has two absolute maxima, lying at the points between the apices of the cusps of the half-moon-shaped ridge and the closest points of the acorn-shaped ridge.

### 5.2.2 *Spatial and Angular Rainbows*

Figure 5.12 shows the spatial and angular rainbow lines in the IP plane for the straight nanotube, for $\alpha_b = 0$. These lines do not cross each other. This means that, for $\alpha_b = 0$, there are no points in the IP plane for which the Jacobians of the mappings of the IP plane to the final TP plane and the TA plane, defined by Eqs. (3.32) and (3.33), both vanish at the exit from the nanotube.

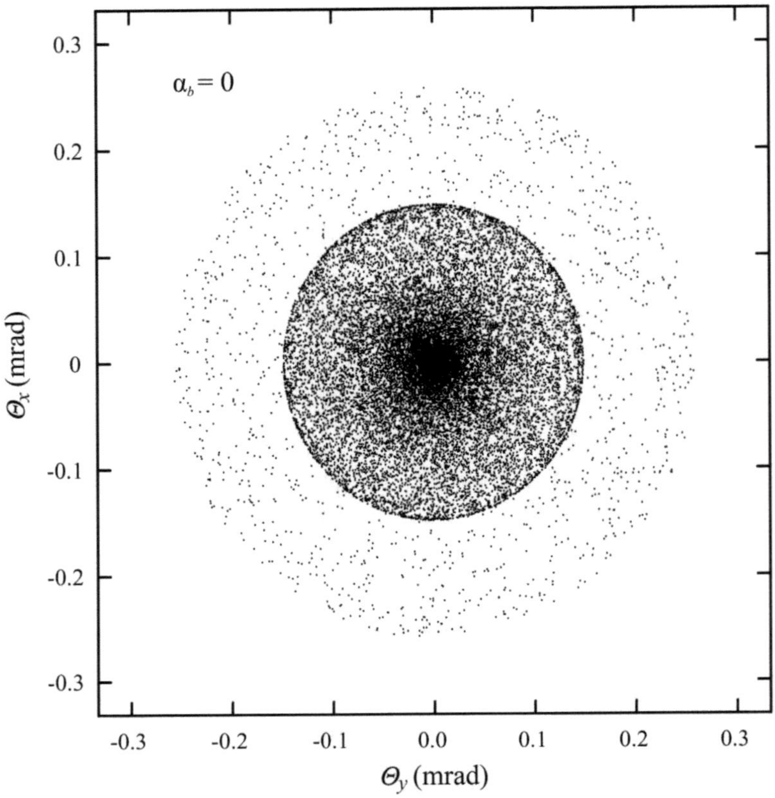


**Fig. 5.9** Angular distribution of 1 GeV protons transmitted through a straight (11, 9) chiral single-wall carbon nanotube for $L = 7$ μm [206]

The rainbow lines in the final TP plane and the TA plane for the straight nanotube, for $\alpha_b = 0$, are given in Fig. 5.13. It is evident that the former line, having the radius of 0.17 nm, coincides with the circular ridge appearing in the corresponding spatial distribution of transmitted protons, which is shown in Fig. 5.6. This means that this circular ridge is a spatial rainbow ridge, which explains the fact that the proton yield changes abruptly across it, i.e., the transmitted proton beam is well confined within it. One can also see that the latter line, having the radius of 0.15 mrad, coincides with the circular ridge occurring in the corresponding angular distribution of transmitted proton, which is given in Fig. 5.9. Hence, this circular ridge is an angular rainbow ridge, which explains the fact that the transmitted proton beam is confined within it. One can conclude that the rainbow effect makes the transmitted proton beam confined in both the final TP plane and the TA plane.

Figure 5.14 gives the spatial rainbow pattern in the IP plane for the bent nanotube when $\alpha_b = 0.2$ mrad. It contains two closed lines and one line open toward the outer nanotube wall. They are designated by $1_s$, $2_s$, and $3_s$. One can

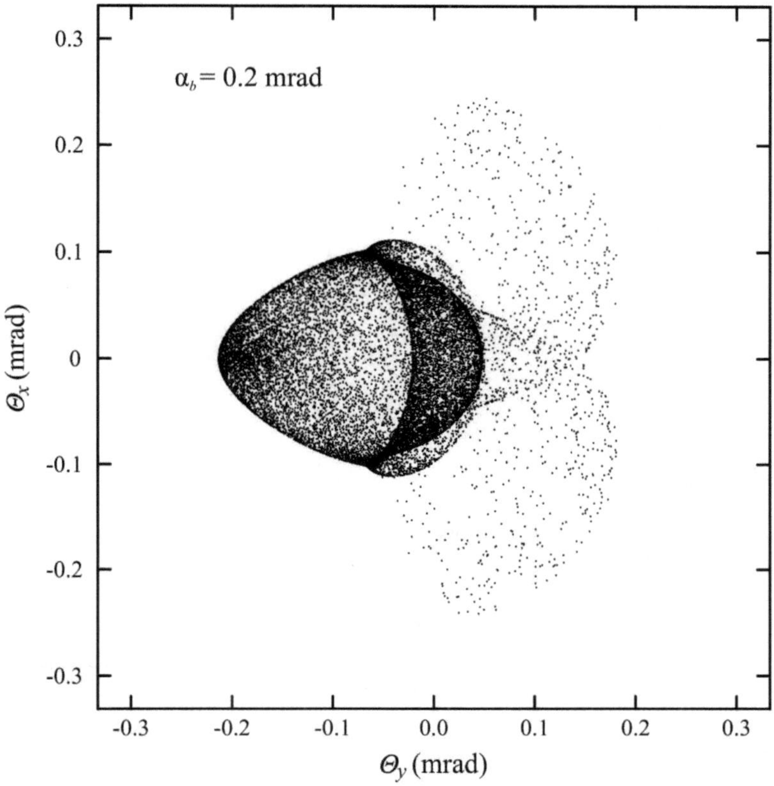


**Fig. 5.10** Angular distribution of 1 GeV protons transmitted through a bent (11, 9) chiral single-wall carbon nanotube for $L = 7$ μm and $\alpha_b = 0.2$ mrad [206]

say that "the center of gravity" of this rainbow pattern is closer to the inner than to the outer nanotube wall. The angular rainbow pattern in the IP plane for the bent nanotube when $\alpha_b = 0.2$ mrad is shown in Fig. 5.15. It contains two closed lines and one line open toward the inner and outer nanotube walls. They are designated by $1_a$, $2_a$, and $3_a$. It is evident that "the center of gravity" of this rainbow pattern is closer to the outer than to the inner nanotube wall. Comparison of these two rainbow patterns shows that they have six crossing points, which are the points in the IP plane for which the Jacobians of the mappings of the IP plane to the final TP plane and the TA plane [defined by Eqs. (3.32) and (3.33)] both vanish at the exit from the nanotube. However, the analysis shows that the images of these points in the final TP plane and the TA plane do not represent any special points in these planes.

The rainbow pattern in the final TP plane for the bent nanotube when $\alpha_b = 0.2$ mrad is given in Fig. 5.16. It contains a quasi-ellipsoidal line, a cusped quasi-rhombic line, and a quasi-circular line open toward the inner nanotube wall. They are the images of lines $1_s$, $2_s$, and $3_s$, and are designated by $1_s'$, $2_s'$, and $3_s'$,

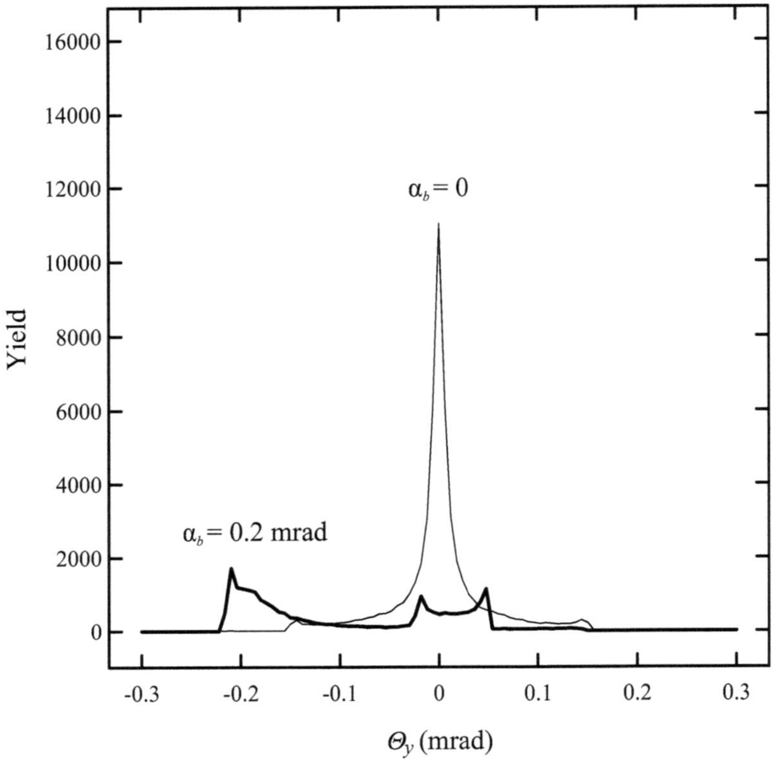


**Fig. 5.11** Yield along the $\Theta_y$ axis of 1 GeV protons transmitted through a straight (11, 9) chiral single-wall carbon nanotube for $L = 7$ μm (*thin line*) and the corresponding yield of protons transmitted through the bent nanotube for $\alpha_b = 0.2$ mrad (*thick line*) [206]

respectively. Lines $1_s'$ and $2_s'$ cross each other at two points. These are the points where the rainbow effect in the final TP plane is enhanced. One can easily see that this rainbow pattern fully explains the corresponding spatial distribution of transmitted protons, which is shown in Fig. 5.7. The transmitted proton beam is defined by lines $1_s'$ and $2_s'$. The rainbow pattern in the TA plane for the bent nanotube when $\alpha_b = 0.2$ mrad is given in Fig. 5.17. It contains a half-moon-shaped line, an acorn-shaped line, and a cusped line open toward the inner and outer nanotube walls. They are the images of lines $1_a$, $2_a$, and $3_a$, and are designated by $1_a'$, $2_a'$, and $3_a'$, respectively. Line $3_a'$ crosses line $2_a'$ at two points, and line $1_a'$ at four points. These are the points at which the rainbow effect in the TA plane is enhanced. It is evident that this rainbow pattern fully explains the corresponding angular distribution of transmitted protons, which is shown in Fig. 5.10. The bent component of the transmitted proton beam is defined by line $1_a'$, the parts of line $2_a'$ close to line $1_a'$, and the parts of line $3_a'$ within line $2_a'$. The above-mentioned absolute maxima of the corresponding angular distribution are connected to the apices of the cusps of lines $1_a'$ and $3_a'$ and the closest points of line $2_a'$.

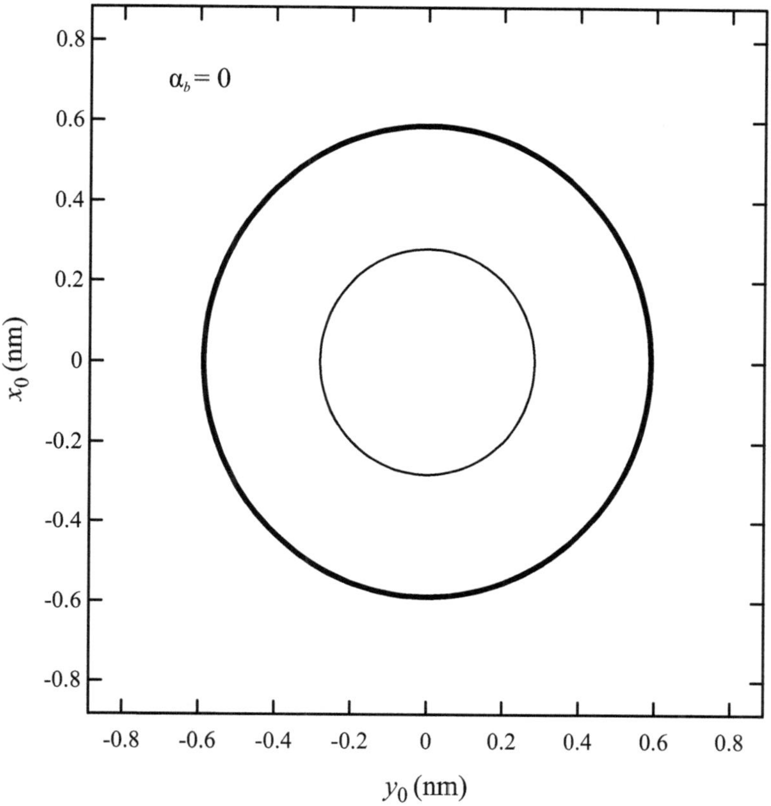


**Fig. 5.12** Spatial and angular rainbow lines in the IP plane for 1 GeV protons transmitted through a straight (11, 9) chiral single-wall carbon nanotube for $L = 7$ μm (*thin* and *thick lines*, respectively) [206]

Figure 5.18 gives the yield of protons transmitted through the bent nanotube for $\alpha_b$ between 0 and 2 mrad normalized to the yield for $\alpha_b = 0$. The incident number of protons is 78,516. A similar dependence was obtained earlier by Zhevago and Glebov [180]. One can take that the normalized proton yield falls at 0.5 for $\alpha_b = \psi_T$ [180]. Consequently, $\psi_T = 0.41$ mrad, and, in accordance with the above-given relation connecting $\psi_c$ and $\psi_T$ [180], $\psi_c = 0.28$ mrad. This means that $\psi_c$ is somewhat larger than the limiting transmission angle of the angular distribution of transmitted protons shown in Fig. 5.9, which equals 0.26 mrad. Hence, the minimal proton distance from the nanotube wall for which its motion is stable is somewhat smaller than $a_{TF}$. It is interesting to compare the bending power of the nanotube to the bending power of a dipole magnet [180]. Since, according to the above given relations, $E_c = 87$ V/nm, the magnetic induction of the dipole magnet having the same bending power as the nanotube for $\alpha_b = \psi_T$ would be $B_c = E_c/v_0 = 330$ T. The conclusion is that the bending power of the nanotube is about 40 times higher than the bending power of the superconducting dipole magnets

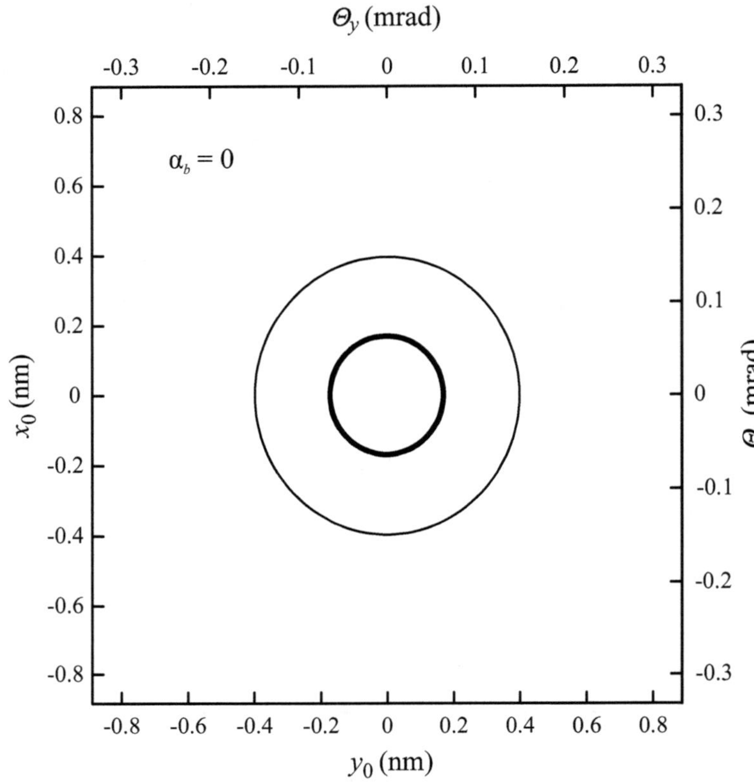


**Fig. 5.13** Rainbow lines in the final TP plane and the TA plane for 1 GeV protons transmitted through a straight (11, 9) chiral single-wall carbon nanotube for $L = 7$ μm (*thin* and *thick lines*, respectively) [206]

within one of the most powerful accelerator in the world, the Large Hadron Collider (LHC), in the European Organization for Nuclear Research (CERN), Geneva, Switzerland [213]. Hence, in principle, it is possible that very small and cheap passive elements composed of bundles of bent (parallel) carbon nanotubes substitute very big and expensive dipole magnets within high-energy accelerator facilities [180].

## 5.3 Spatial and Angular Rainbows with Straight Very Long Nanotubes

Let us now consider a 1 GeV proton moving through a very long (11, 9) chiral single-wall carbon nanotube [201, 214]. The nanotube length, $L$, is varied from 10 to 500 μm. The $z$ axis of the reference frame coincides with the nanotube axis,

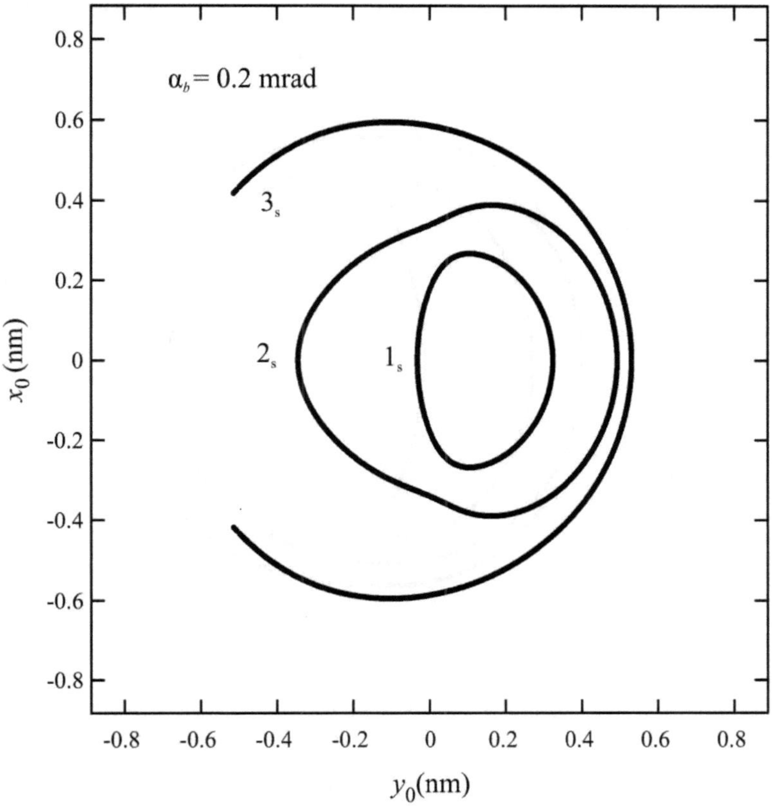


**Fig. 5.14** Spatial rainbow lines in the IP plane for 1 GeV protons transmitted through a bent (11, 9) chiral single-wall carbon nanotube for $L = 7$ μm and $\alpha_b = 0.2$ mrad [206]

and its origin lies in the entrance plane of the nanotube. The $x$ and $y$ axes of the reference frame are the vertical and horizontal axes, respectively. The incident proton velocity vectors are parallel to the nanotube axis.

In the calculations, we employ the theory of crystal rainbows (Sect. 3.4). The interaction of the proton and a nanotube atom is described by Molière's approximation of the Thomas-Fermi interaction potential given by Eqs. (2.1), (2.2), and (2.4), with $Z_1 = 1$ and $Z_2 = 6$. We employ the continuum approximation as in Subsect. 5.2.1 but with the thermal vibrations of the nanotube atoms taken into account. In a way analogous to the one giving Eq. (3.13), Eq. (5.4) and (5.5) lead to the thermally averaged proton-nanotube continuum interaction potential

$$V_{\text{ch}}^{\text{th}}(x, y) = V_{\text{ch},0} \sum_{\ell=1}^{3} \left( \alpha_\ell + \frac{\sigma_{\text{th}}^2 \beta_\ell^2}{2a_{\text{TF}}^2} \right) K_0 \left( \frac{\beta_\ell R_n}{a_{\text{TF}}} \right) I_0 \left( \frac{\beta_\ell \rho}{a_{\text{TF}}} \right) \quad (5.7)$$

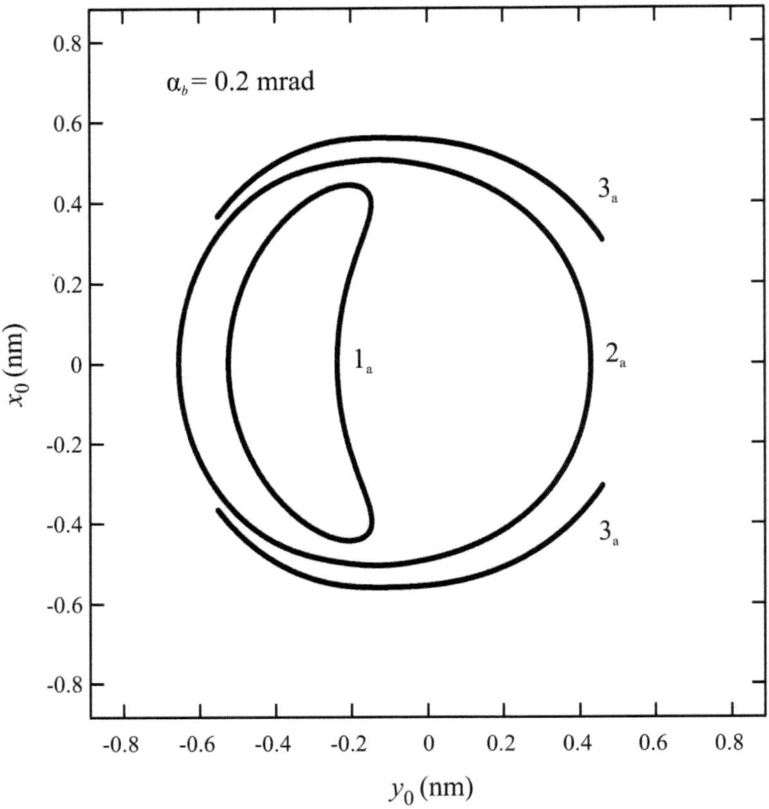


**Fig. 5.15** Angular rainbow lines in the IP plane for 1 GeV protons transmitted through a bent (11, 9) chiral single-wall carbon nanotube for $L = 7$ μm and $\alpha_b = 0.2$ mrad [206]

[132], where $\sigma_{\text{th}} = 5.3$ pm is the one-dimensional atomic thermal vibration amplitude [212]. It is evident that it is cylindrically symmetric. The electronic proton energy loss and the uncertainty of the proton channeling angle due to its collisions with the nanotube electrons are disregarded. The components of the proton impact parameter vector are chosen randomly from the uniform distribution within a circle around the origin of the radius $R_n - a_{\text{TF}}$.

Since $V_{\text{ch}}^{\text{th}}(x, y)$ is cylindrically symmetric, the problem we consider is in fact one-dimensional. This means that the rainbow lines in the IP plane and the TA plane, if they exist, will show up as circles. Consequently, it is sufficient to analyze the mapping of the $x_0$axis in the IP plane to the $X$ axis in the final TP plane, $X(x_0)$, being the spatial proton transmission function, and the mapping of the $x_0$axis to the $\Theta_x$ axis in the TA plane, $\Theta_x(x_0)$, being the angular proton transmission function. The abscissas and ordinates of the extrema of $X(x_0)$ determine the radii of the spatial rainbow lines in the IP plane and the rainbow lines in the final TP plane, while the abscissas and ordinates of the extrema of $\Theta_x(x_0)$ determine the radii of the angular

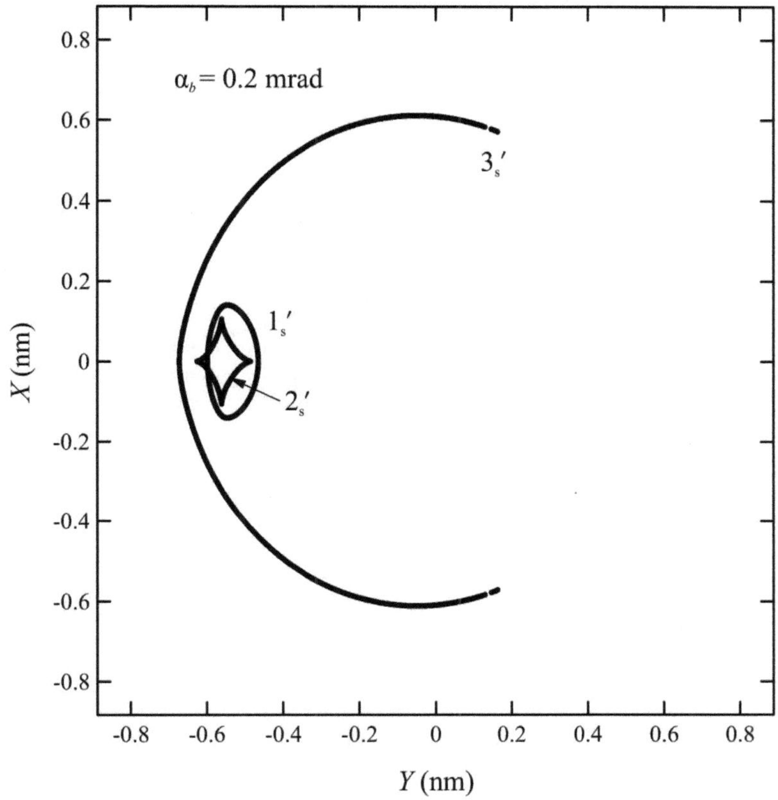


**Fig. 5.16** Rainbow pattern in the final TP plane for 1 GeV protons transmitted through a bent (11, 9) chiral single-wall carbon nanotube for $L = 7$ μm and $\alpha_b = 0.2$ mrad [206]

rainbow lines in the IP plane and the rainbow lines in the TA plane, respectively [21].

In the following two subsections, we shall explore in detail how (i) the axial and zero-degree yields of protons transmitted through the nanotube, (ii) the spatial and angular distributions of transmitted protons, (iii) the spatial and angular proton transmission functions, and (iv) the final proton positions and channeling angles change with $L$. This will enable one to envision and fully understand the proton-nanotube interaction in question in the corresponding four-dimensional phase space. We would like to mention a study by Berry and O'Dell [215] of rays propagation over long distances through a sinusoidal volume grating, with the refractive index varying perpendicularly to the direction of incidence. We have not been aware of that study until recently, and the results and conclusions to be presented here have not been influenced in any way by the results and conclusions given in it. However, one can find some qualitative similarities between them.

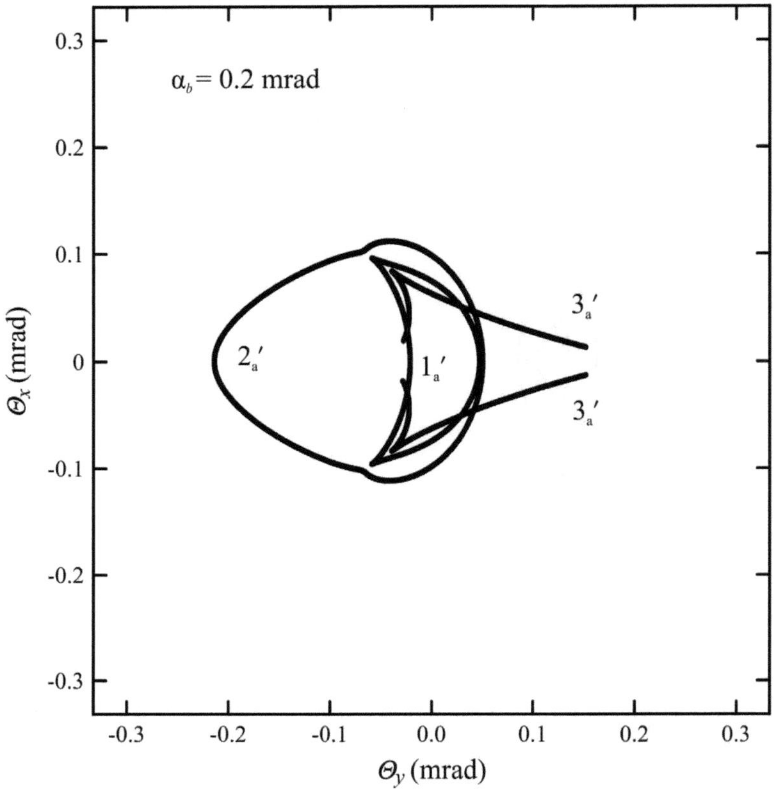


**Fig. 5.17** Rainbow pattern in the TA plane for 1 GeV protons transmitted through a bent (11, 9) chiral single-wall carbon nanotube for $L$ = 7 μm and $\alpha_b = 0.2$ mrad [206]

### 5.3.1 Spatial Rainbow Effect

Figure 5.19a depicts the normalized axial yield of transmitted protons, i.e., their yield around the origin in the final TP plane divided by the incident number of protons, $Y_s$, as a function of $L$. The incident number of protons is 314,132,833. For the region in the final TP plane around the origin, we take the region in which the radial component of the proton position vector, $\mathrm{P} = (X^2 + Y^2)^{1/2}$, where $X$ and $Y$ are its vertical and horizontal components, respectively, is less or equal to $0.025R_n$, where $R_n = 0.69$ nm is the nanotube radius [172]. The dependence has 13 maxima, which correspond to the superfocusing effect, appearing at the middle points of the first 13 rainbow cycles (see Sect. 4.3).

Figures 5.20a, b show the spatial distributions of transmitted protons along the $X$ axis for $L$ = 10 and 100 μm, respectively [214]. The size of a bin along the $X$ axis is 0.0028 nm, and the incident number of protons is the same as in the case of the axial yield of transmitted protons. The spatial distributions contain a central maximum, and 2 and 15 symmetric pairs of maxima characterized by a

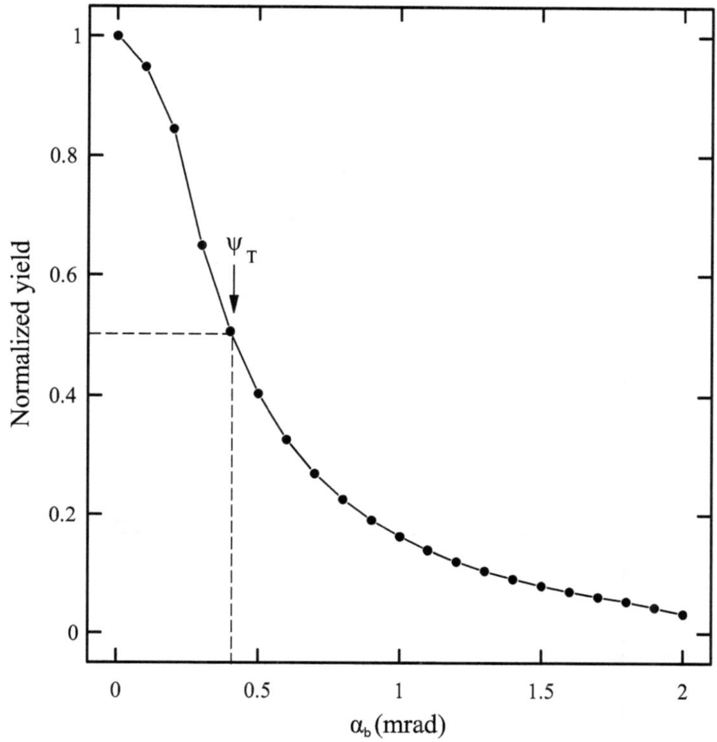


**Fig. 5.18** Normalized yield of 1 GeV protons transmitted through a bent (11, 9) chiral single-wall carbon nanotube for $L = 7$ μm and $\alpha_b$ between 0 and 2 mrad [206]

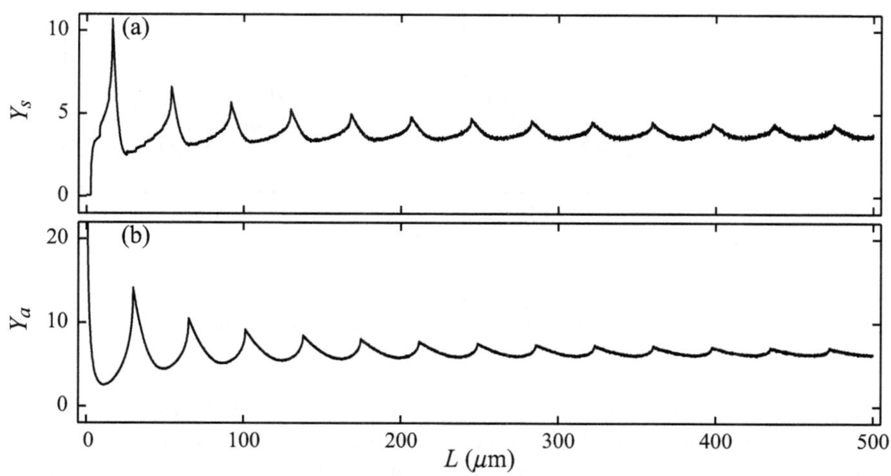


**Fig. 5.19** (**a**) Normalized axial yield of 1 GeV protons transmitted through a (11, 9) chiral single-wall carbon nanotube as a function of $L$. (**b**) Corresponding normalized zero-degree yield of protons transmitted through the nanotube as a function of $L$

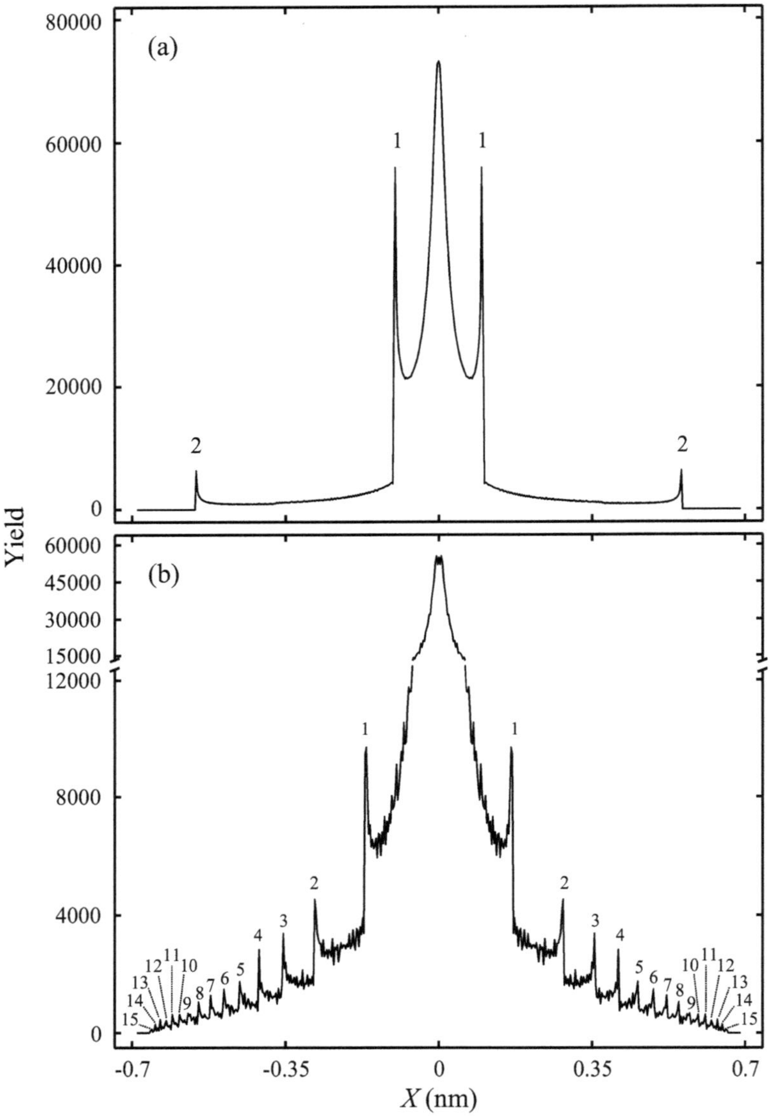


**Fig. 5.20** Spatial distributions along the *X* axis of 1 GeV protons transmitted through a (11, 9) chiral single-wall carbon nanotube of lengths of (**a**) 10 μm and (**b**) 100 μm

sharp decrease of the proton yield on the large position side, respectively. The spatial proton transmission functions obtained for $L = 10$ and 100 μm are given in Figs. 5.21a, b, respectively [214]. They contain 2 and 15 symmetric pairs of extrema, respectively, with each pair including a minimum and maximum. Each of them defines a circular rainbow line in the IP plane and a circular rainbow line in the final TP plane. Comparison of Figs. 5.20 and 5.21 demonstrates that the

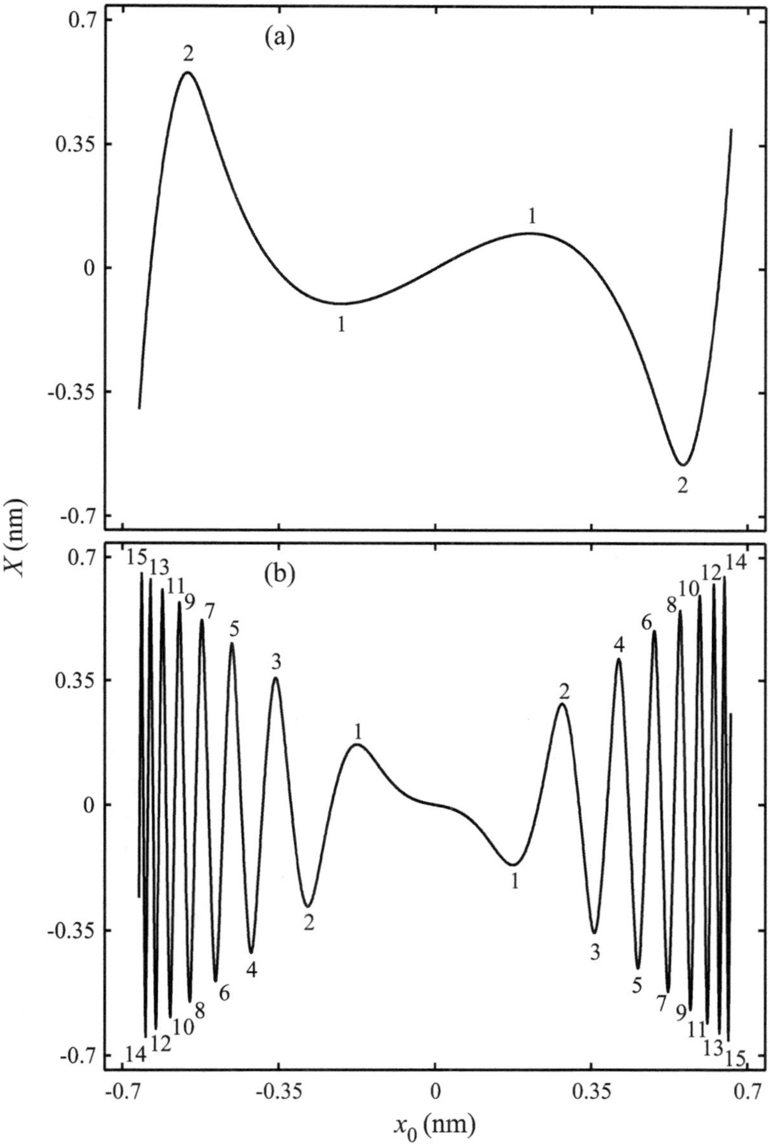


**Fig. 5.21** Spatial transmission functions of 1 GeV protons transmitted through a (11, 9) chiral single-wall carbon nanotube of lengths of (**a**) $L = 10$ μm and (**b**) $L = 100$ μm

abscissas of the pairs of sharp maxima of each spatial distribution coincide with the ordinates of the corresponding pairs of extrema of the associated spatial transmission function. Thus, the sharp maxima of the spatial distributions can be attributed to the spatial rainbow effect. The bright and dark sides of each spatial rainbow are its inner and outer sides, respectively.

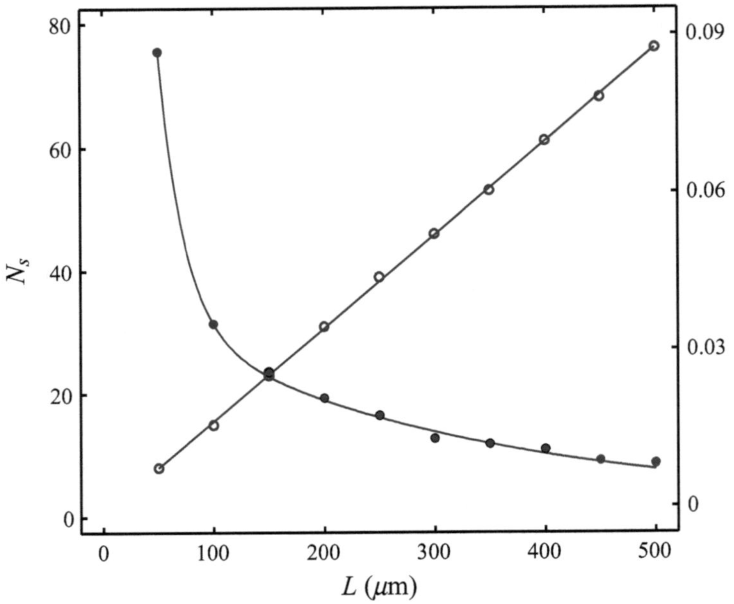


**Fig. 5.22** Dependences of the number of spatial rainbows occurring with 1 GeV protons and a (11, 9) chiral single-wall carbon nanotube (*open circles*) and the average distance between them (*closed circles*) on $L$

The spatial distributions of transmitted protons along the $X$ axis and the associated spatial transmission functions in the whole investigated region of $L$, between 10 and 500 μm, show that the number of obtained spatial rainbows increases and the average distance between them decreases as $L$ increases. These two dependences, for $L$ between 50 and 500 μm, are given in Fig. 5.22 [214]. The former dependence increases linearly, while the latter one decreases exponentially as $L$ increases. For a sufficiently long nanotube, the average distance between the spatial rainbows becomes smaller than the resolution of the spatial distribution, and one cannot distinguish between the adjacent rainbows. Then, the spatial rainbows disappear and the spatial distribution becomes a bell-shaped one.

### 5.3.2 Angular Rainbow Effect

The normalized zero-degree yield of transmitted protons, i.e., their yield around the origin in the TA plane divided by the incident number of protons, $Y_a$, as a function of $L$ is given in Fig. 5.19b. The incident number of protons is again 314,132,833. For the region in the TA plane around the origin, we take the region in which the proton transmission angle, $\Theta = \left(\Theta_x^2 + \Theta_y^2\right)^{1/2}$, where $\Theta_x$ and $\Theta_y$ are its vertical and horizontal components, respectively, is smaller than $0.025\psi_c$, where

$\psi_c = 0.23$ mrad is the critical angle for axial channeling [see Eq. (2.12)]. The dependence has 13 maxima, which correspond to the zero-degree focusing effect, occurring at the end points of the first 13 rainbow cycles (see Sect. 3.3).

Figures 5.23a, b show the angular distributions of transmitted protons along the $\Theta_x$ axis for $L$ = 10 and 100 μm, respectively [201]. The size of a bin along the

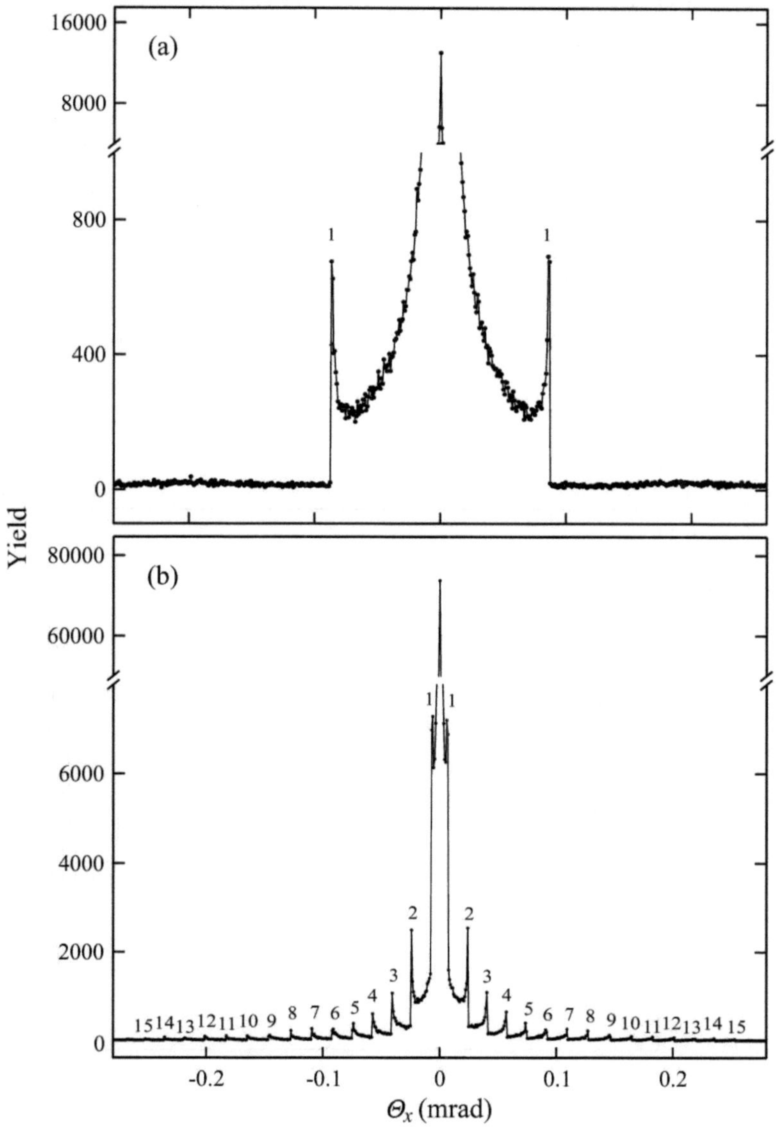


**Fig. 5.23** Angular distributions along the $\Theta_x$ axis of 1 GeV protons transmitted through a (11, 9) chiral single-wall carbon nanotube of lengths of (**a**) 10 μm and (**b**) 100 μm [201]

$\Theta_x$ axis is 0.866 μrad, and the incident number of protons is 16,656,140. The angular distributions contain a central maximum, and 1 and 15 symmetric pairs of maxima characterized by a sharp decrease of the proton yield on the large angle side, respectively. The angular proton transmission functions obtained for $L = 10$ and 100 μm are given in Figs. 5.24a, b, respectively [201]. They contain 1 and

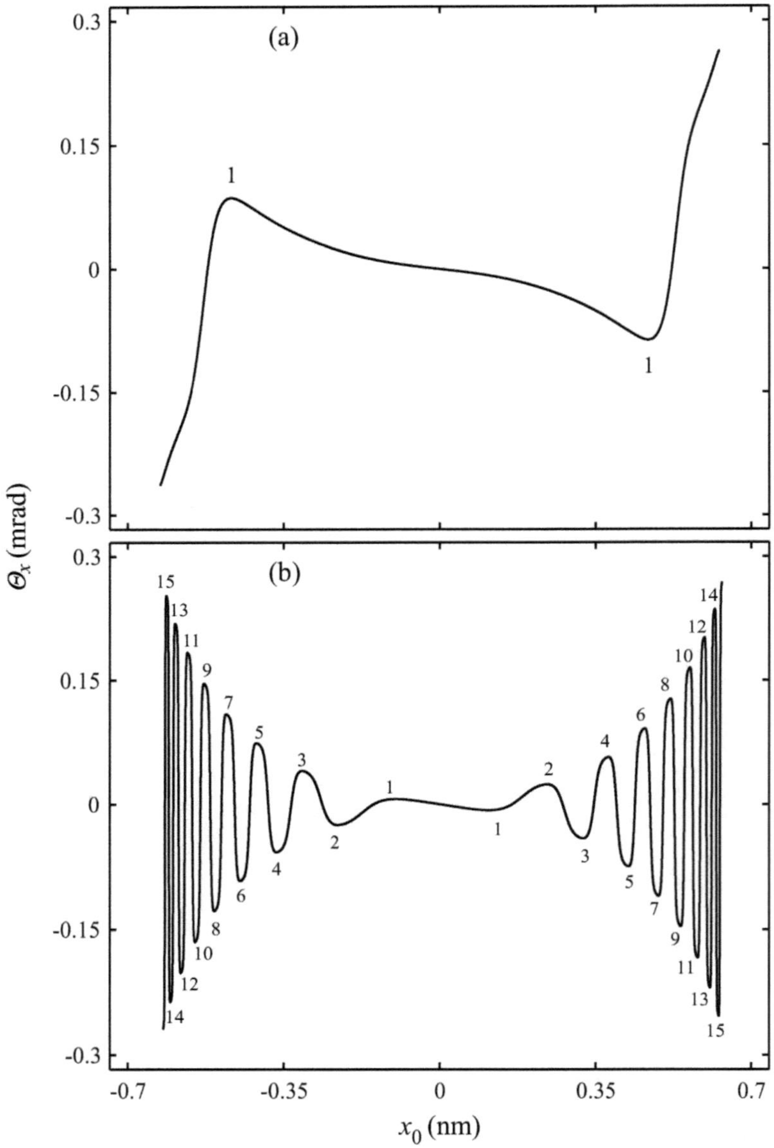


**Fig. 5.24** Angular transmission functions of 1 GeV protons transmitted through a (11, 9) chiral single-wall carbon nanotube of lengths of (**a**) $L = 10$ μm and (**b**) $L = 100$ μm [201]

15 symmetric pairs of extrema, respectively, with each pair including a minimum and maximum. Each of them defines a circular rainbow line in the IP plane and a circular rainbow line in the TA plane. Comparison of Figs. 5.23 and 5.24 shows that the abscissas of the pairs of sharp maxima of each angular distribution coincide with the ordinates of the corresponding pairs of extrema of the associated angular transmission function. Hence, the sharp maxima of the angular distributions can be explained by the angular rainbow effect. The bright and dark sides of the angular rainbows are its inner and outer sides, respectively.

Analysis of the angular distributions of transmitted protons along the $\Theta_x$ axis and the associated angular transmission functions in the whole investigated region of $L$, between 10 and 500 μm, demonstrates that the number of generated angular rainbows increases and the average distance between them decreases as $L$ increases. These two dependences, for $L$ between 50 and 500 μm, are given in Fig. 5.25 [201]. The former dependence increases linearly, while the latter one decreases exponentially as $L$ increases. When the nanotube becomes sufficiently long for the average distance between the angular rainbows to become smaller than the resolution of the angular distribution, one cannot distinguish between the adjacent rainbows. Consequently, the angular rainbows disappear and the angular distribution becomes a bell-shaped one. Since, as it has been shown in the previous subsection, the same happens to the spatial distribution of transmitted protons along the $x$ axis, one can say that, when the nanotube becomes sufficiently long, the

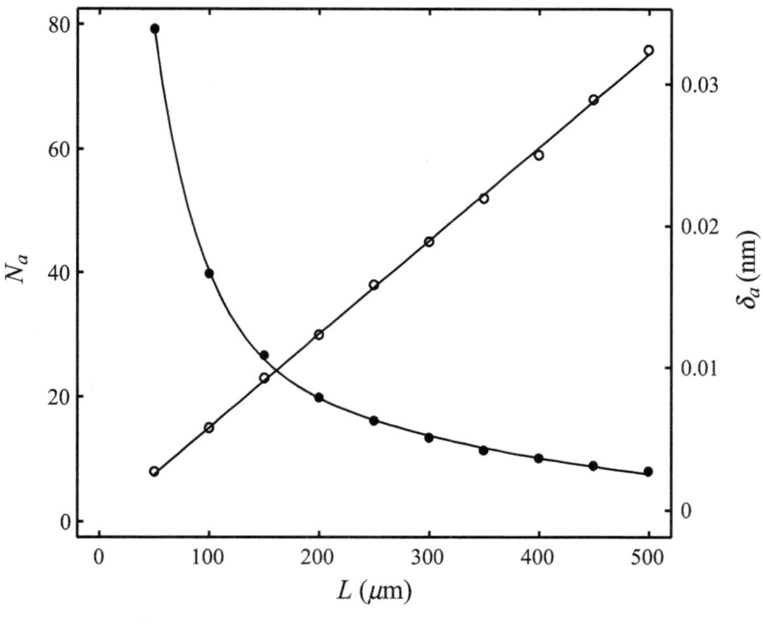


**Fig. 5.25** Dependences of the number of angular rainbows occurring with 1 GeV protons and a (11, 9) chiral single-wall carbon nanotube (*open circles*) and the average distance between them (*closed circles*) on $L$ [201]

angular and spatial distributions equilibrate. As one would expect, this does not happen in accordance with the ergodic hypothesis in ion channeling [67]. We call this route to equilibration, which is characterized by the linear increases of the numbers of spatial and angular rainbows and the exponential decreases of the distances between them as $L$ increases, the rainbow route to equilibration [201].

### 5.3.3 Trajectories

But how does the trajectories of the protons propagating along the nanotube look like? How do the final proton positions and channeling angles depend on $L$? Figure 5.26a gives the final proton positions, $X$, as functions of $L$ in the region comprising the first superfocusing point for 100 values of $x_0$ between $-(R_n - a_{TF})$ and $R_n - a_{TF}$ with the step of $\Delta x_0 = 0.0133$ nm. The nanotube length corresponding to the first superfocusing point, where the reduced nanotube length equals 0.250, is $L_1^{sf} = 19.6$ μm. One can see that the lines representing the functions $X(L; x_0)$ in the considered region of $L$ have four envelopes. The first envelope is defined by the proton trajectories in the first half of the first rainbow cycle, and it terminates at the first superfocusing point. For each value of $L$, there are two points belonging to each envelope. Analysis shows that these points are the spatial rainbow points. This means that they belong to a spatial rainbow line in the final TP plane, and that the envelope belongs to a spatial rainbow surface, which terminates at the first superfocusing point. This cusped surface may be referred to as the first spatial rainbow surface. Similarly, the second, third, and fourth envelopes, which are defined by the proton trajectories extending down to the lines parallel to the $X$ axis determined by the second, third, and fourth superfocusing points, not seen in the figure, belong to the second, third, and fourth spatial rainbow surfaces, respectively. Figure 5.26b depicts all the envelopes in the considered region of $L$. This region comprises the first 13 superfocusing points, where the first 13 spatial rainbow surfaces terminate. The number of intersections of the line parallel to the $X$ axis determined by a value of $L$ with the envelopes divided by 2 equals the number of spatial rainbow lines in the final TP plane appearing for this value of $L$ (see Fig. 5.20).

The final proton channeling angles, $\Theta_x$, as functions of $L$ in the region comprising the first zero-degree focusing point for 100 values of $x_0$ between $-(R_n - a_{TF})$ and $R_n - a_{TF}$ with the step of $\Delta x_0$=0.0133 nm are shown in Fig. 5.27a. The nanotube length corresponding to the first zero-degree focusing point, where the reduced nanotube length equals 0.500, is $L_1^{zdf} = 2L_1^{sf} = 39.2$ μm. One can see that the lines representing the functions $\Theta_x(L; x_0)$ in the region of $L$ under consideration have six envelopes. The first envelope is defined by the proton trajectories in the first rainbow cycle, and it terminates at the first zero-degree focusing point.

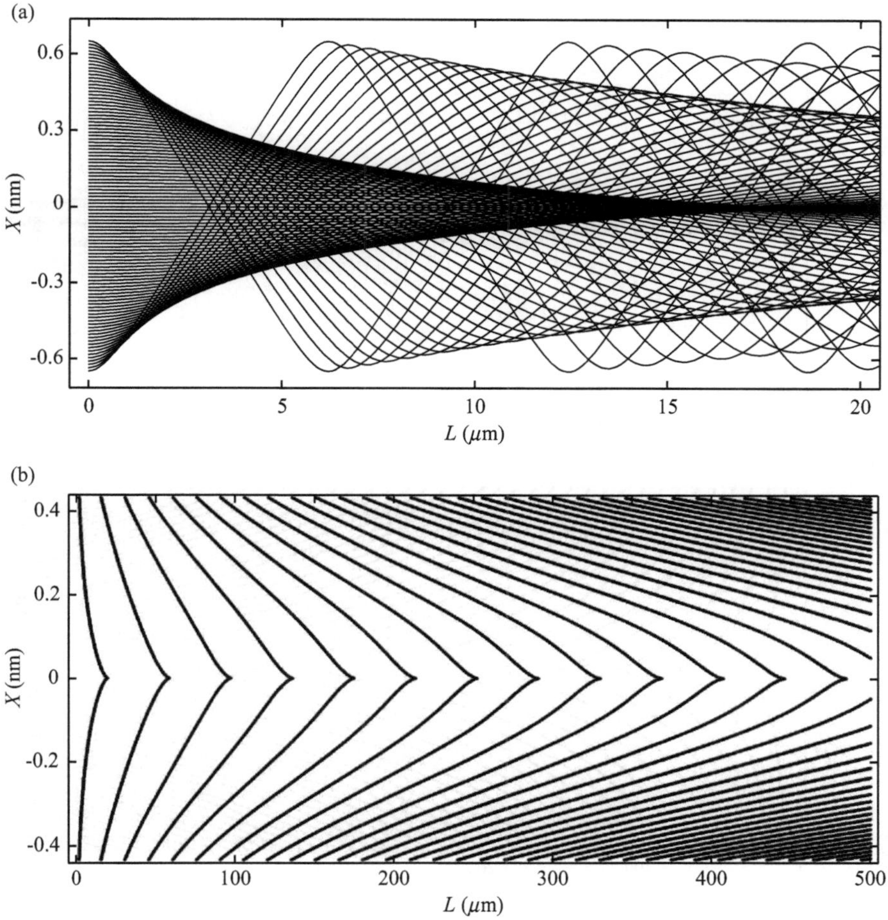


**Fig. 5.26** (**a**) Final proton positions as functions of $L$ in the region comprising the first superfocusing point in the case of 1 GeV protons and a (11, 9) chiral single-wall carbon nanotube. (**b**) Envelopes of the *lines* representing the functions $X(L; x_0)$ for $L$ between 10 and 500 μm

For each value of $L$, there are two points belonging to each envelope. These points are the angular rainbow points. Hence, they belong to an angular rainbow line in the TA plane, and the envelope belongs to an angular rainbow surface, which terminates at the first zero-degree focusing point. This cusped surface may be referred to as the first angular rainbow surface. In a similar way, the second, third, etc. envelopes, which are defined by the proton trajectories extending down to the lines parallel to the $\Theta_x$ axis determined by the second, third, etc. superfocusing points, not seen in the figure, belong to the second, third, etc. angular rainbow surfaces, respectively. Figure 5.27b shows all the pairs of envelopes in the region of $L$ we explore. This region comprises the first 13 zero-degree focusing points, where

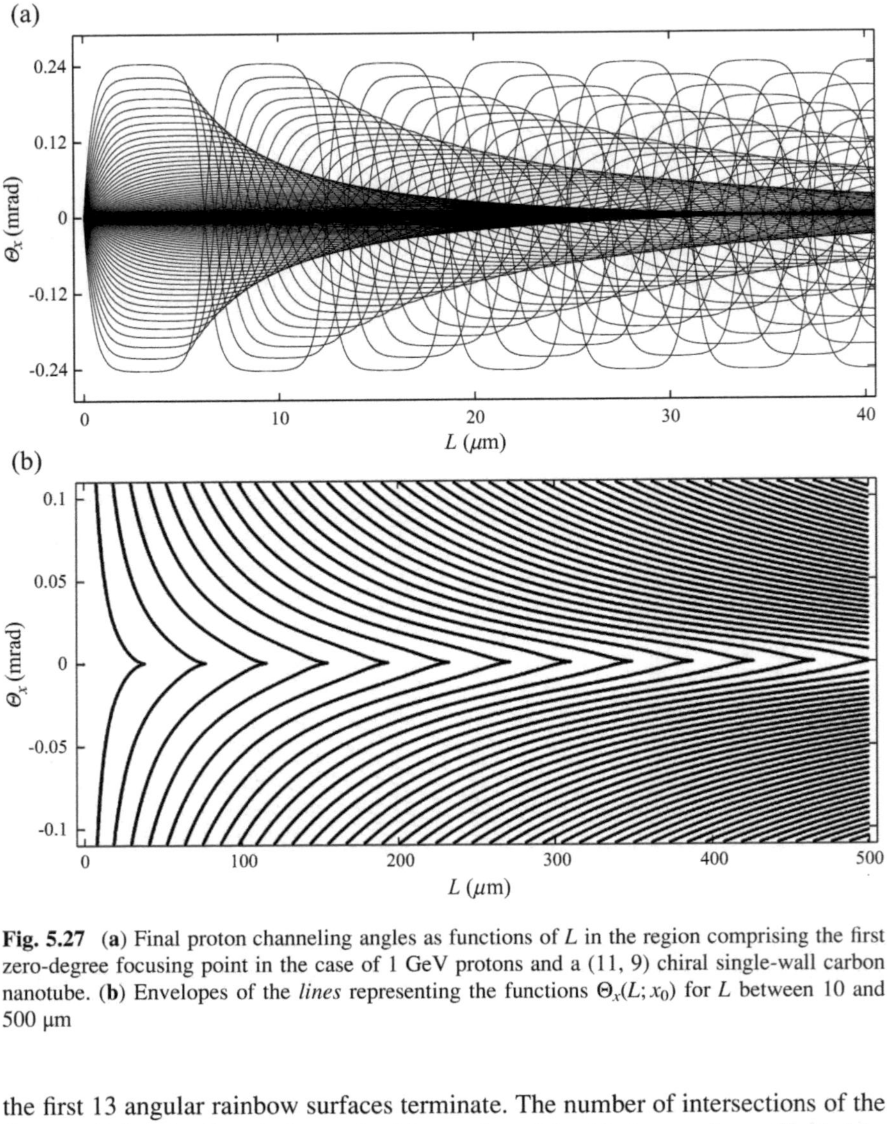


**Fig. 5.27** (**a**) Final proton channeling angles as functions of $L$ in the region comprising the first zero-degree focusing point in the case of 1 GeV protons and a (11, 9) chiral single-wall carbon nanotube. (**b**) Envelopes of the *lines* representing the functions $\Theta_x(L; x_0)$ for $L$ between 10 and 500 μm

the first 13 angular rainbow surfaces terminate. The number of intersections of the line parallel to the $\Theta_x$ axis determined by a value of $L$ with the envelopes divided by 2 is equal to the number of angular rainbow lines in the TA plane occurring for this value of $L$ (see Fig. 5.23).

# Chapter 6
# Rainbows with Positrons and Carbon Nanotubes

This chapter is devoted to the rainbows occurring in channeling of positrons of incident kinetic energy of $E_0 = 1$ MeV in (11, 9) chiral single-wall carbon nanotubes. As it has been said and explained in Subsect. 2.3.1, the process will be treated using quantum mechanics. In this case, the mass of the projectiles is sufficiently small and their incident kinetic energy sufficiently low that they clearly exhibit their wave properties. As it has been demonstrated in Chap. 1, in such a case, each rainbow is composed of a principal rainbow and one or more supernumerary rainbows. The principal rainbow is the primary, secondary, or a higher order rainbow.

Andersen et al. [216] registered the quantum behavior of 1 MeV positrons channeled in a gold crystal and then backscattered from it. A similar experiment with the same result was performed with 1.2 MeV positrons and a silicon crystal by Pedersen et al. [217]. In both measurements, the positrons moved along the planar crystal channels. Haakwnaasen et al. [218] observed the pronounced quantum behavior of 1 MeV positrons transmitted through the axial and planar channels of a silicon crystal. All those measurements were analyzed using the dynamical diffraction theory, which is based on the expansion of the positron wave function in the Bloch wave functions [219].

Recently, Petrović et al. [50] presented a classical and quantum mechanical investigation of transmission of positrons through very short (11, 9) carbon nanotubes. The incident positron kinetic energies were 1 and 10 MeV, while the nanotube lengths were 200 and 560 nm, respectively. The analysis was focused on the rainbow effects, which were clearly observed in the spatial and angular distributions of transmitted positrons. The quantum rainbow maxima were explained by the constructive interference of two rays within the wave packet that started from the points in the IP plane on two sides of the rainbow point and finished at the same point in the final TP plane or the TA plane. Those authors used the same approach to explore the possibility of using the quantum rainbows for characterization of very short carbon nanotubes [220].

 
N. Nešković et al., *Rainbows in Channeling of Charged Particles in Crystals and Nanotubes*, Lecture Notes in Nanoscale Science and Technology 25,
DOI 10.1007/978-3-319-61524-0_6

Here, we shall begin with a presentation of the quantum mechanical theory of rainbow channeling [50]. Then, the classical and quantum spatial and angular rainbows generated with 1 MeV positrons and very short (11, 9) carbon nanotubes will be considered [221]. The nanotube length, $L$, will be changed from 50 to 320 nm.

We take that the $z$ axis of the reference frame, being the longitudinal axis, coincides with the nanotube axis and that its origin lies in the nanotube entrance plane. The $x$ and $y$ axes, being the transverse axes, are the vertical and horizontal axes, respectively. Since the nanotubes are very short, the positron energy losses caused by its collisions with the nanotube nuclei and electrons as well as the uncertainty of the positron channeling angle due to its collisions with the nanotube electrons are all neglected.

## 6.1 Quantum Mechanical Theory of Rainbow Channeling

### 6.1.1 Interaction Potentials

In this study, the interaction of the positron and a nanotube atom is described by Molière's approximation of the Thomas-Fermi interaction potential given by Eqs. (2.1), (2.2), and (2.4), with $Z_1=1$ and $Z_2=6$. In a similar way as in Sect. 5.3, the needed continuum interaction potential of the positron and nanotube is obtained in three steps. The first step is the axial averaging of the positron-atom interaction potential on the basis of the continuum approximation (Subsect. 2.3.2). The second step is the azimuthal averaging of the interaction potential obtained in the first step on the basis of the fact that the nanotube is chiral. The resulting positron-nanotube continuum interaction potential reads

$$V_{\mathrm{ch}}(\rho; R_n) = V_{\mathrm{ch},0} \sum_{\ell=1}^{3} \alpha_\ell K_0\left(\frac{\beta_\ell R_n}{a_{\mathrm{TF}}}\right) I_0\left(\frac{\beta_\ell \rho}{a_{\mathrm{TF}}}\right) \quad \text{for } \rho \leq R_n \tag{6.1}$$

and

$$V_{\mathrm{ch}}(\rho; R_n) = V_{\mathrm{ch},0} \sum_{\ell=1}^{3} \alpha_\ell I_0\left(\frac{\beta_\ell R_n}{a_{\mathrm{TF}}}\right) K_0\left(\frac{\beta_\ell \rho}{a_{\mathrm{TF}}}\right) \quad \text{for } \rho > R_n \tag{6.2}$$

with

$$V_{\mathrm{ch},0} = \frac{16\pi Z_2 e^2 R_{\mathrm{n}}}{3^{4/3} a_b^2}, \tag{6.3}$$

where $R_n$, $a_b$, $\rho$, $x$, $y$, $I_0$, and $K_0$ are the same as in Eqs. (5.4) and (5.5) [132]. It should be mentioned that Eqs. (6.1) and (6.3) coincide with Eqs. (5.4) and (5.5), respectively. The function defined by Eqs. (6.1), (6.2), and (6.3) is continuous, but its first derivative is discontinuous at the point $\rho = R_n$.

In the third step, the thermal vibrations of the nanotube atoms are introduced. This is done by the initial averaging of the interaction potential of the positron and a nanotube atom over its displacements from the equilibrium position along the $x$, $y$, and $z$ axes [78]. These displacements are taken to be small and independent, and are described by a Gaussian distribution function. This averaging appears in the third step as the averaging of the interaction potential obtained in the second step over the effective thermally induced changes of the nanotube radius, $R$, from its equilibrium radius, $R_n$, along the $\rho$ axis. The resulting positron-nanotube continuum interaction potential is given by

$$V_{\mathrm{qu}}^{\mathrm{th}}(\rho;R_n) = \frac{1}{(2\pi)^{1/2}\sigma_{\mathrm{th}}}\int_{R_1}^{R_2} V_{\mathrm{ch}}(\rho;R)\exp\left[-\frac{(R-R_n)^2}{2\sigma_{\mathrm{th}}^2}\right]\mathrm{d}R, \tag{6.4}$$

where $\sigma_{\mathrm{th}} = 5.3$ pm is the one-dimensional atomic thermal vibration amplitude [212]. The integration limits appearing in this expression are $R_1 = R_n - 6\sigma_{\mathrm{th}}$ and $R_2 = R_n + 6\sigma_{\mathrm{th}}$. They have been chosen to comprise the interval in which the changes of the integrand are not negligible. The integration is performed numerically [128]. This function and its first derivative are continuous, i.e., its first derivative is not discontinuous at the point $\rho = R_n$. However, since the displacements of the nanotube atoms are small, a useful analytical approximation of Eq. (6.4) can be obtained by substituting $V_{\mathrm{ch}}(\rho, R)$, represented as a function of $\rho - R$, with its Taylor series about the point $\rho - R_n$ truncated after its second-order term [201]. For $\rho \leq R_n$, this expression becomes

$$V_{\mathrm{cl}}^{\mathrm{th}}(\rho;R_n) = V_{\mathrm{ch},0}\sum_{\ell=1}^{3}\left(\alpha_\ell + \frac{\sigma_{\mathrm{th}}^2\beta_\ell^2}{2a_{\mathrm{TF}}^2}\right)K_0\left(\frac{\beta_\ell R_n}{a_{\mathrm{TF}}}\right)I_0\left(\frac{\beta_\ell\rho}{a_{\mathrm{TF}}}\right), \tag{6.5}$$

coinciding with Eq. (5.7). This function is continuous, but its first derivative is discontinuous at the point $\rho = R_n$, in a similar way as $V_{\mathrm{ch}}(\rho; R_n)$, given by Eqs. (6.1) and (6.3). Analysis has shown that the difference between $V_{\mathrm{cl}}^{\mathrm{th}}(\rho;R_n)$ and $V_{\mathrm{qu}}^{\mathrm{th}}(\rho;R_n)$ is very small, especially for $\rho \leq R_n - a_{\mathrm{TF}}$.

In the classical calculations to be presented here, we have used $V_{\mathrm{cl}}^{\mathrm{th}}(\rho;R_n)$, given by Eqs. (6.3) and (6.5), as the needed positron-nanotube continuum interaction potential. This has been done because in these calculations $\rho \leq R_n - a_{\mathrm{TF}}$. On the other hand, the quantum mechanical calculations have been performed with $V_{\mathrm{qu}}^{\mathrm{th}}(\rho;R_n)$, given by Eqs. (6.1), (6.2), (6.3), and (6.4), as the needed continuum interaction potential. The reason is the fact that the discontinuity of the function $V_{\mathrm{cl}}^{\mathrm{th}}(\rho;R_n)$ at the point $\rho = R_n$, which must not be excluded from these calculations, has proven to be a source of serious numerical problems. It is important to note that the functions $V_{\mathrm{cl}}^{\mathrm{th}}(\rho; R_n)$ and $V_{\mathrm{qu}}^{\mathrm{th}}(\rho; R_n)$ are cylindrically symmetric.

We would like to emphasize that $V_{\mathrm{th}}^{\mathrm{qu}}(\rho; R_n)$ represents a circular potential barrier of a finite height, which means that the positron always tunnels through the nanotube wall, i.e., it is always partly dechanneled. However, it has been found that when the nanotube is very short, the positron dechanneling effect is much less pronounced than the channeling effect [221].

### 6.1.2 Classical Approach

The vertical and horizontal components of the initial positron position vector, in the IP plane, are denoted as $x_0$ and $y_0$, and the vertical and horizontal components of its initial momentum vector as $p_{x0}$ and $p_{y0}$, respectively. Further, the vertical and horizontal components of the positron position vector during the channeling are denoted as $x(t)$ and $y(t)$, and the vertical and horizontal components of its momentum vector during the channeling as $p_x(t)$ and $p_y(t)$, respectively, where $t$ denotes time. In the classical calculations, we solve the equations of motion. Since the positron-atom interaction potential is axially averaged, $x(t)$, $y(t)$, $p_x(t)$, and $p_y(t)$ are obtained via the equations of motion in the transverse plane and the positron-nanotube interaction potential is determined by Eqs. (6.3) and (6.5). These equations read

$$m_e d_{tt} x = -\partial_x V_{\mathrm{cl}}^{\mathrm{th}}(\rho; R_n) \tag{6.6}$$

and

$$m_e d_{tt} y = -\partial_y V_{\mathrm{cl}}^{\mathrm{th}}(\rho; R_n), \tag{6.7}$$

where $m_e$ is the positron mass and $d_{tt} \equiv d^2/dt^2$. They are solved numerically [128]. In the case under consideration, the critical angle for axial channeling is calculated by the expression

$$\psi_c = \left[\frac{V_{\mathrm{cl}}^{\mathrm{th}}(R_n - a_{\mathrm{TF}}; R_n)}{E_0}\right]^{1/2} \tag{6.8}$$

[66], giving $\psi_c = 7.34\,\mathrm{mrad}$. This means that the angle between the positron momentum vector and nanotube axis is always smaller than $\psi_c$. Therefore, the vertical and horizontal components of the positron channeling angle, in the channeling angle (CA) plane, can be taken to be $\theta_x(t) = p_x(t)/p$ and $\theta_y(t) = p_y(t)/p$, respectively, where $p$ is the magnitude of the positron momentum vector. Since the positron energy losses during the channeling are disregarded, $p$ equals the

magnitude of its incident momentum vector, $p_0$. The initial values of these functions are $\theta_{x0} = p_{x0}/p_0$ and $\theta_{y0} = p_{y0}/p_0$, respectively.

It has been said above that the positron-nanotube continuum interaction potential is cylindrically symmetric. Consequently, as in the cases described in Sect. 5.3, the mapping of the IP plane to the TP plane or the CA plane is reduced to the mapping of, e.g., the $x_0$ axis in the IP plane to the $x$ axis in the TP plane or the $\theta_x$ axis in the CA plane. The former mapping, $x(x_0, t)$, is the positron spatial channeling function, while the latter mapping, $\theta_x(x_0, t)$, is the positron angular channeling function. One can also analyze the diagrams $\tilde{x}(\theta_x, t) = -x[x_0(\theta_x), t]$ and $\widetilde{\theta}_x(x, t) = \theta_x[x_0(x), t]$ [215, 222]. The former diagram is the negative inverse of the latter one and vice versa. These diagrams are referred to as the spatial and angular rainbow diagrams of the positrons channeled in the nanotube, respectively.

Let us now introduce the classical spatial and angular Hamilton principal functions of the positrons channeled in the nanotube, $S_\rho^{\mathrm{cl}}(x, t)$ and $S_\theta^{\mathrm{cl}}(\theta_x, t)$, respectively. In accordance with a description given in Ref. [223], these functions satisfy the equations

$$\frac{\mathrm{d}S_\rho^{\mathrm{cl}}(x, t)}{\mathrm{d}x} = p\widetilde{\theta}_x(x, t) \tag{6.9}$$

and

$$\frac{\mathrm{d}S_\theta^{\mathrm{cl}}(\theta_x, t)}{\mathrm{d}\theta_x} = p\tilde{x}(\theta_x, t). \tag{6.10}$$

These equations cannot be solved directly since the functions on their right sides are multivalued. However, they can be transformed into the equations

$$\frac{\mathrm{d}\tilde{S}_\rho^{\mathrm{cl}}(x_0, t)}{\mathrm{d}x_0} = p\theta_x(x_0, t)\frac{\mathrm{d}x(x_0, t)}{\mathrm{d}x_0} \tag{6.11}$$

and

$$\frac{\mathrm{d}\tilde{S}_\theta^{\mathrm{cl}}(x_0, t)}{\mathrm{d}x_0} = -px(x_0, t)\frac{\mathrm{d}\theta_x(x_0, t)}{\mathrm{d}x_0}, \tag{6.12}$$

respectively, where $\tilde{S}_\rho^{\mathrm{cl}}(x_0, t) = S_\rho^{\mathrm{cl}}[x(x_0), t]$ and $\tilde{S}_\theta^{\mathrm{cl}}(x_0, t) = S_\theta^{\mathrm{cl}}[\theta_x(x_0), t]$. The functions on the right sides of these equations are single-valued. The equations are solved numerically [128]. Their solutions lead directly to $S_\rho^{\mathrm{cl}}(x, t)$ and $S_\theta^{\mathrm{cl}}(\theta_x, t)$, respectively.

### 6.1.3 Quantum Mechanical Approach

In the quantum mechanical calculations, the approach is via the time-dependent Schrödinger equation. Since the positron-atom interaction potential is axially averaged, we solve the time-dependent Schrödinger equation in the transverse plane, which is

$$i\hbar\partial_t\psi_s(x,y,t) = -\frac{\hbar^2}{2m_e}\Delta\psi_s(x,y,t) + V_{\mathrm{qu}}^{\mathrm{th}}(\rho;R_n)\psi_s(x,y,t), \quad (6.13)$$

where $\psi_s(x, y, t)$ is the positron wave function in the TP plane, $\partial_t \equiv \partial/\partial t$, and $\hbar$ is the reduced Planck constant.

The incident positron beam is represented as an ensemble of noninteracting Gaussian wave packets [50]. As a result, we implicitly take into account the divergence of the incident beam relative to the nanotube axis. It should be emphasized that such a treatment differs from the previous treatments of positron channeling in crystals [216–218], in which the dynamical diffraction theory was used.

An incident wave packet in the spatial representation is described by

$$\psi_s(x,y,t=0;x_0,y_0) = \frac{1}{(2\pi)^{1/2}\sigma_\rho}\exp\left[-\frac{(x-x_0)^2+(y-y_0)^2}{4\sigma_\rho^2}\right], \quad (6.14)$$

where $\sigma_\rho$ is its spatial standard deviation. On the other hand, the incident wave packet in the angular representation is defined by

$$\psi_a(\theta_x,\theta_y,t=0;x_0,y_0) = \frac{1}{(2\pi)^{1/2}\sigma_\theta}\exp\left[-\frac{ip(x_0\theta_x+y_0\theta_y)}{\hbar}\right]\exp\left(-\frac{\theta_x^2+\theta_y^2}{4\sigma_\theta^2}\right), \quad (6.15)$$

which is obtained from the Fourier transform of Eq. (6.14), where $\sigma_\theta = \Delta_d/(8\ln 2)^{1/2}$ is its angular standard deviation with $\Delta_d$ being the incident beam angular FWHM. The parameters $\sigma_\rho$ and $\sigma_\theta$ are connected to each other with the corresponding uncertainty principle, resulting in the expression $\sigma_\rho\sigma_\theta = \hbar/(2p)$.

The final wave packet in the spatial representation is defined by the function $\psi_s(x, y, t_f; x_0, y_0)$ with $t_f = m_e L/p$, which is obtained via Eq. (6.13). The equation is solved numerically [224]. We take that the computation domain is the region $-D_x/2 \le x \le D_x/2$ and $-D_y/2 \le y \le D_y/2$ with $D_x = D_y = 3R_n$. The spatial distribution of transmitted positrons is given as

$$Y_s(x,y) = \sum_{x_0,y_0} c(x_0,y_0)|\psi_s(x,y,t_f;x_0,y_0)|^2, \quad (6.16)$$

which is a weighted sum of the final positron spatial probability functions over the chosen values of $x_0$ and $y_0$, with the pairs of values of $x_0$ and $y_0$ chosen within the region $\rho \leq R_n - a_{TF}$. This should be done randomly from the uniform distribution, and the number of these pairs of values should be large. In that case, each of the coefficients $c(x_0, y_0)$ would be equal to $1/N_p$, with $N_p$ being the incident number of positrons. However, in order to minimize the computation time, the pairs of values of $x_0$ and $y_0$ are chosen using an algorithm providing a minimal deviation of the incident spatial distribution of positrons from the uniform distribution within the region $\rho \leq R_n - a_{TF}$ with a smaller number of these pairs. This uniform distribution is given by $1/[\pi(R_n - a_{TF})^2]$. We achieve this by adjusting the values of $c(x_0, y_0)$.

The final wave packet in the angular representation is described by the function $\psi_a(\theta_x, \theta_y, t_f; x_0, y_0)$, which is the positron wave function in the CA plane, obtained from the Fourier transform of $\psi_s(x, y, t_f; x_0, y_0)$. The transformation is performed numerically [128]. The angular distribution of transmitted positrons is given as

$$Y_a(\theta_x, \theta_y) = \sum_{x_0, y_0} c(x_0, y_0)|\psi_a(\theta_x, \theta_y, t_f; x_0, y_0)|^2, \tag{6.17}$$

which is the weighted sum of the final positron angular probability functions over the chosen values of $x_0$ and $y_0$. It should be noted that $Y_s(x, y)$ and $Y_a(\theta_x, \theta_y)$, which correspond to the whole ensemble of positrons, are cylindrically symmetric.

The exponential form of the function $\psi_s(x, y, t)$ is

$$\psi_s(x, y, t) = A_\rho(x, y, t)\exp[iS_\rho^{\text{qu}}(x, y, t)/\hbar], \tag{6.18}$$

where $A_\rho(x, y, t)$ and $S_\rho^{\text{qu}}(x, y, t)/\hbar$ are the amplitude and phase of this function, respectively, determined by its real and imaginary parts. Similarly, the exponential form of the function $\psi_a(\theta_x, \theta_y, t)$ is

$$\psi_a(\theta_x, \theta_y, t) = A_\theta(\theta_x, \theta_y, t)\exp[iS_\theta^{\text{qu}}(\theta_x, \theta_y, t)/\hbar], \tag{6.19}$$

where $A_\theta(\theta_x, \theta_y, t)$ and $S_\theta^{\text{qu}}(\theta_x, \theta_y, t)/\hbar$ are the amplitude and phase of this function, respectively, determined by its real and imaginary parts. In accordance with a consideration contained in Ref. [225], $S_\rho^{\text{qu}}(x, y, t)$ and $S_\theta^{\text{qu}}(\theta_x, \theta_y, t)$ are recognized as the quantum spatial and angular Hamilton principal functions of a positron channeled in the nanotube, respectively. One should note that these functions, which correspond to one positron from the ensemble, are not cylindrically symmetric.

## 6.2 Spatial and Angular Primary Rainbows

### 6.2.1 *Classical Rainbows*

We have calculated and inspected the angular rainbow diagrams of the positrons transmitted through the nanotube, $\widetilde{\theta}_x(x, t_f)$, for nanotube lengths $L = 50$–$320$ nm, with a step of 10 nm [221]. This means that the reduced nanotube length, $\Lambda$, has been varied between 0.03 and 0.19. Figure 6.1 gives $\widetilde{\theta}_x(x, t_f)$ for $L = 320$ nm. The

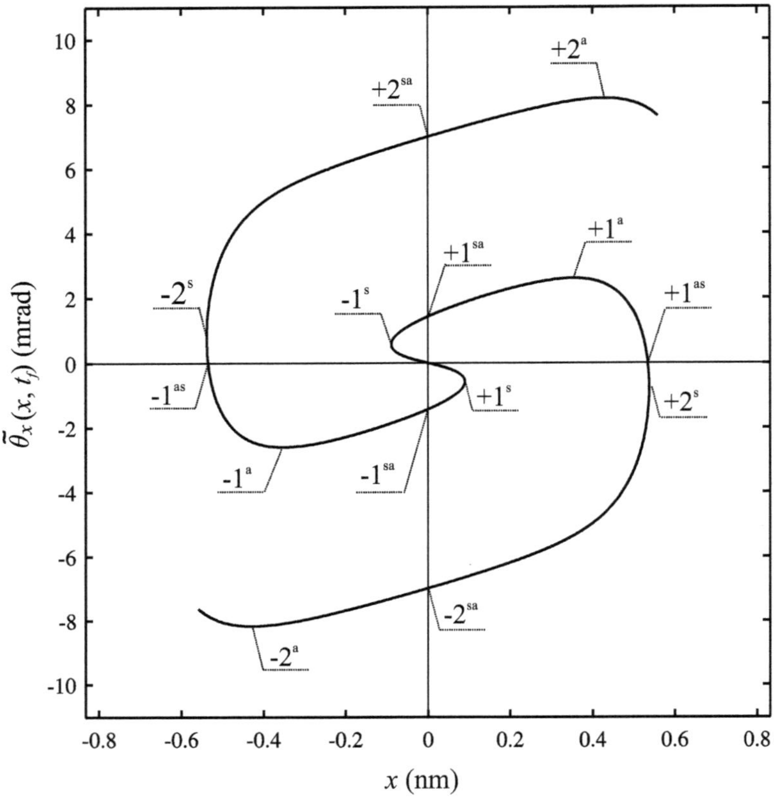


**Fig. 6.1** Angular rainbow diagram of 1 MeV positrons transmitted through a (11, 9) chiral single-wall carbon nanotube for $L = 320$ nm and $p_{x0} = p_{y0} = 0$. Points $\pm 1^s$ and $\pm 2^s$ belong to the classical SPR and SSR, and $\pm 1^a$ and $\pm 2^a$ to the classical APR and ASR, respectively [221]. Points $\pm 1^{sa}$ and $\pm 2^{sa}$ designate the intersections of the curve with the $\tilde{\theta}_x$ axis off the origin and $\pm 1^{as}$ its intersections with the $x$ axis off the origin

incident positron momentum vector is taken to be parallel to the nanotube axis, i.e., $p_{x0} = p_{y0} = 0$. This curve also represents the associated spatial rainbow diagram, $\tilde{x}(\theta_x, t)$. The positions of the rainbow extrema in the final TP plane are determined by the tangents to the curve perpendicular to the $x$ axis, while the positions of the rainbow extrema in the TA plane are determined by the tangents to it perpendicular to the $\tilde{\theta}_x$ axis [50, 215]. In this case, there are two pairs of extrema in the final TP plane, designated as $\pm 1^s$ and $\pm 2^s$, each of which represents a maximum and minimum of the positron spatial transmission function, $x(x_0, t_f)$, respectively. Hence, there are two circular rainbow lines in the final TP plane. Analysis shows that the positron trajectories that correspond to extrema $\pm 1^s$ involve one deflection from the nanotube wall and the trajectories that correspond to extrema $\pm 2^s$ two deflections from the wall. Consequently, the former extrema is associated with a classical spatial primary rainbow (SPR), and the latter extrema with a classical spatial secondary rainbow (SSR).

Besides, there are two pairs of extrema in the TA plane, designated as $\pm 1^a$ and $\pm 2^a$, each of which represents a maximum and minimum of the positron angular transmission function, $\theta_x(x_0, t_f)$, respectively. Thus, there are two circular rainbow lines in the TA plane. The former extrema is associated with a classical angular primary rainbow (APR), and the latter extrema with a classical angular secondary rainbow (ASR).

One should observe that the curve in question intersects the $\widetilde{\theta}_x$ axis at the origin, at a pair of points closer to the origin, $\pm 1^{sa}$, and at a pair of points farther from the origin, $\pm 2^{sa}$. Each point from the former pair lies between the points corresponding to the SPR and APR, and each point from the latter pair between the points corresponding to the SSR and ASR. On the other hand, the curve intersects the $x$ axis at the origin and at a pair of other points, $\pm 1^{as}$. Each point from the pair lies between the points corresponding to the APR and SSR.

Analysis has shown that, in the case under consideration, the angular rainbow diagram of the positrons channeled in the nanotube, $\widetilde{\theta}_x(x, t)$, evolves in the following way (see Fig. 6.1) [215, 221]. The initial curve lies on the $x$ axis. Its evolution begins with its left and right branches starting to bend upward and downward, respectively. Then, the curve reaches points $\pm 1^s$, where the evolution direction changes. The former branch continues rightward and upward, and the latter branch leftward and downward. After that, the curve passes through points $\pm 1^{sa}$ and reaches points $\pm 1^a$, where the evolution direction changes again. The former branch continues rightward and downward, and the latter branch leftward and upward. In the further evolution, it passes through points $\pm 1^{as}$, reaches points $\pm 2^s$, passes through points $\pm 2^{sa}$, and reaches points $\pm 2^a$. This means that the sequence of appearance of the rainbows in this channeling process is: SPR, APR, SSR, ASR, etc.

Figure 6.2 shows the classical spatial Hamilton principal function of transmitted positrons for $L = 320$ nm, $S_\rho^{\mathrm{cl}}(x, t_f)$, for $p_{x0} = p_{y0} = 0$. The curve representing this function has two pairs of cusp singular points, two pairs of inflection points, and two crunode singular points. The cusp singular points closer to the origin correspond to the SPR, $\pm 1^s$; the cusp singular points farther from the origin to the SSR, $\pm 2^s$; the inflection points closer to the origin to the APR, $\pm 1^a$; and the inflection points farther from the origin to the ASR, $\pm 2^a$. The crunode singular points, both lying at the origin, correspond to the two pairs of points where the curve $\widetilde{\theta}_x(x, t_f)$ intersects the $\widetilde{\theta}_x$ axis off the origin, $\pm 1^{sa}$ and $\pm 2^{sa}$ (see Fig. 6.1).

The classical angular Hamilton principal function of transmitted positrons for $L = 320$ nm, $S_\theta^{\mathrm{cl}}(\theta_x, t_f)$, for $p_{x0} = p_{y0} = 0$ is depicted in Fig. 6.3. The curve representing this function has two pairs of cusp singular points, two pairs of inflection points, and a crunode singular point. The cusp singular points closer to the origin correspond to the APR, $\pm 1^a$; the cusp singular points farther from the origin to the ASR, $\pm 2^a$; the inflection points closer to the origin to the SPR, $\pm 1^s$; and the inflection points farther from the origin to the SSR, $\pm 2^s$. The crunode

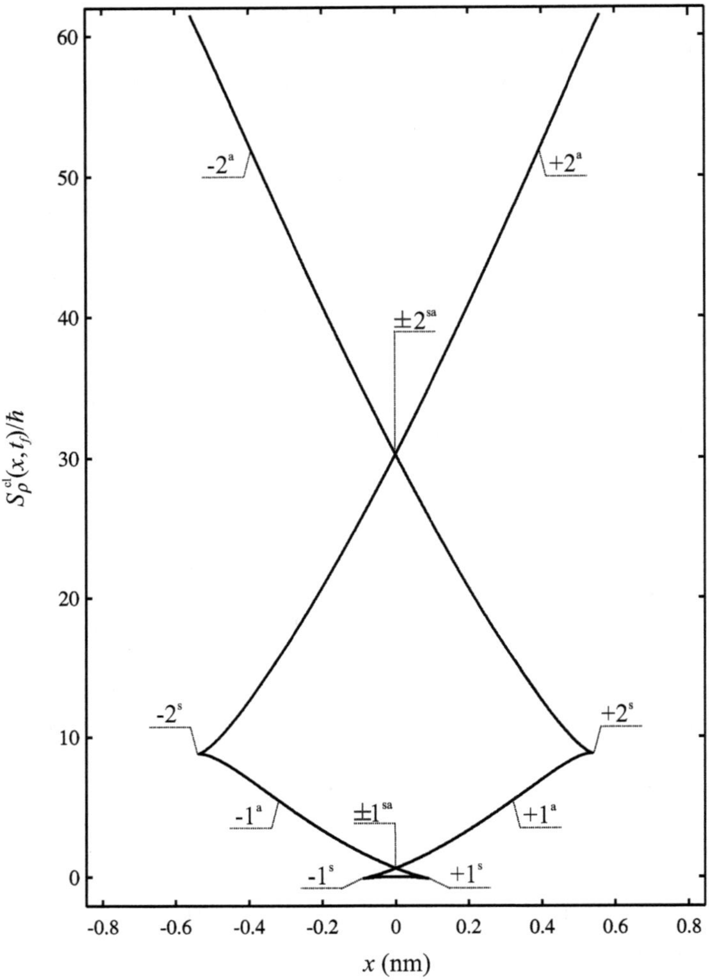


**Fig. 6.2** Classical spatial Hamilton principal function of 1 MeV positrons transmitted through a (11, 9) chiral single-wall carbon nanotube for $L = 320$ nm and $p_{x0}=p_{y0}=0$ divided by $\hbar$ [221]. Points $\pm 1^s$ and $\pm 2^s$ are the cusp singular points of the curve, $\pm 1^a$ and $\pm 2^a$ its inflection points, and $\pm 1^{sa}$ and $\pm 2^{sa}$ its crunode singular points

singular point corresponds to the pair of points where the curve $\widetilde{\theta}_x(x, t_f)$ intersects the $x$ axis off the origin, $\pm 1^{as}$ (see Fig. 6.1).

### 6.2.2 *Quantum Rainbows*

We present here the cases of transmission of the positrons through the nanotubes of lengths of $L = 50$, 100, and 250 nm [221]. The incident positron beam is assumed to

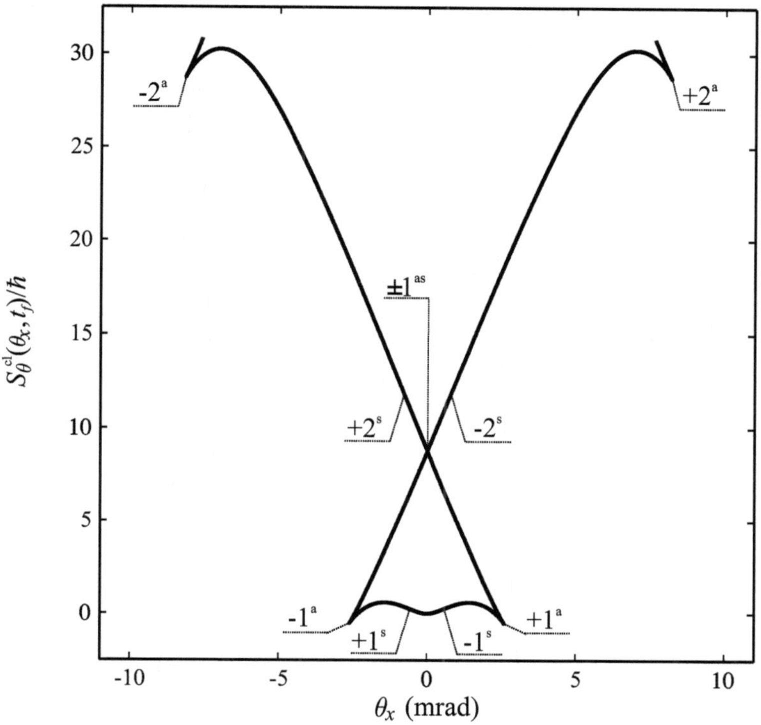


**Fig. 6.3** Classical angular Hamilton principal function of 1 MeV positrons transmitted through a (11, 9) chiral single-wall carbon nanotube for $L = 320$ nm and $p_{x0}=p_{y0}=0$ divided by $\hbar$ [221]. Points $\pm 1^a$ and $\pm 2^a$ are the cusp singular points of the curve, $\pm 1^s$ and $\pm 2^s$ its inflection points, and $\pm 1^{as}$ its crunode singular points

be an ensemble of noninteracting Gaussian wave packets with angular standard deviation of $\sigma_\theta = 0.1\psi_c = 0.73$ mrad, giving a spatial standard deviation of $\sigma_\rho = 0.19$ nm. The incident number of positrons is $N_p$=301. In the cases of $L = 50$ and 100 nm, only the quantum SPR effect is developed, while in the case of $L = 250$ nm, this is true for the SPR, APR, and SSR effects. In the case of $L = 250$ nm, we shall describe only the APR effect.

#### 6.2.2.1 Quantum Spatial Primary Rainbows for $L = 50$ nm

Figure 6.4 shows the one-dimensional normalized spatial distribution of transmitted positrons for $L = 50$ nm. It contains a pair of strong maxima at $\pm 0.40$ nm, and three pairs of weaker maxima at $\pm 0.28$ nm, $\pm 0.20$ nm, and $\pm 0.12$ nm. Let us explain this distribution.

Figure 6.5 gives the positron spatial transmission function, $x(x_0, t_f)$, in the case of $L = 50$ nm. It has a maximum and minimum, $\pm 1^s$, at $x_{01}^s = \pm 0.54$ nm, which are $x_1^s = \pm 0.44$ nm, respectively, belonging to the classical SPR. This means that for $x_0$

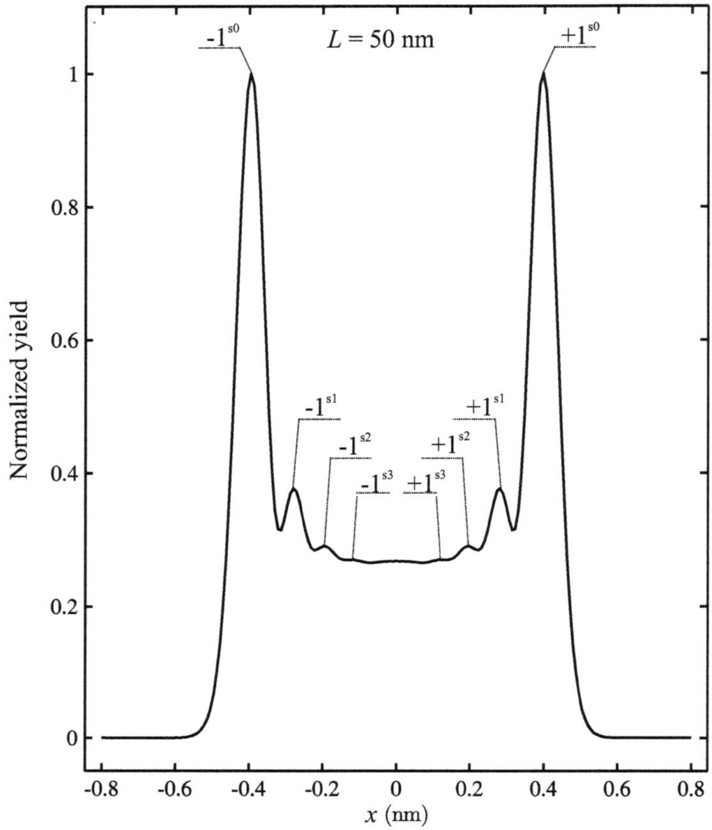


**Fig. 6.4** One-dimensional normalized spatial distribution of 1 MeV positrons transmitted through a $(11, 9)$ chiral single-wall carbon nanotube for $L = 50$ nm described initially as an ensemble of Gaussian wave packets [221]. Maxima $\pm 1^{s0}$ belong to the principal SPR, and maxima $\pm 1^{s1}$, $\pm 1^{s2}$, and $\pm 1^{s3}$ to the supernumerary SPRs

close to $x^s_{01}$, the positrons concentrate in the regions just before $x^s_1$ going from the origin. Also, there are no positrons in the regions after $x^s_1$. Thus, the region before $x^s_1$ is the bright side of the rainbow, and the region after $x^s_1$ its dark side. This figure also contains the two-dimensional representation of the amplitudes squared of the positron wave functions in the final TP plane along the $x$ axis, $A^2_\rho(x, y = 0, t_f)$, for $-R_n < x_0 < R_n$ and $y_0 = 0$, i.e., for $\rho_0 = (x_0^2 + y_0^2)^{1/2} < R_n$ and $\phi_0 = \tan^{-1}(y_0/x_0) =$ 0 and $\pi$, in the case of $L = 50$ nm. This two-dimensional distribution has a pair of maxima, $\pm 1^{m0}$, at $x_0^{m0} = \pm 0.53$ nm and $x^{m0} = \pm 0.41$ nm, respectively. For $x_0$ close to zero, these functions have the Gaussian shape with the width larger but close to the initial width. However, for $x_0$ beyond $\pm 0.26$ nm going from the origin, each of these functions has a few clearly observable maxima or shoulders – it is wrinkled. This is to be attributed to the effect of internal focusing of the corresponding wave packet. For example, a ray within the wave packet entering the nanotube at a smaller

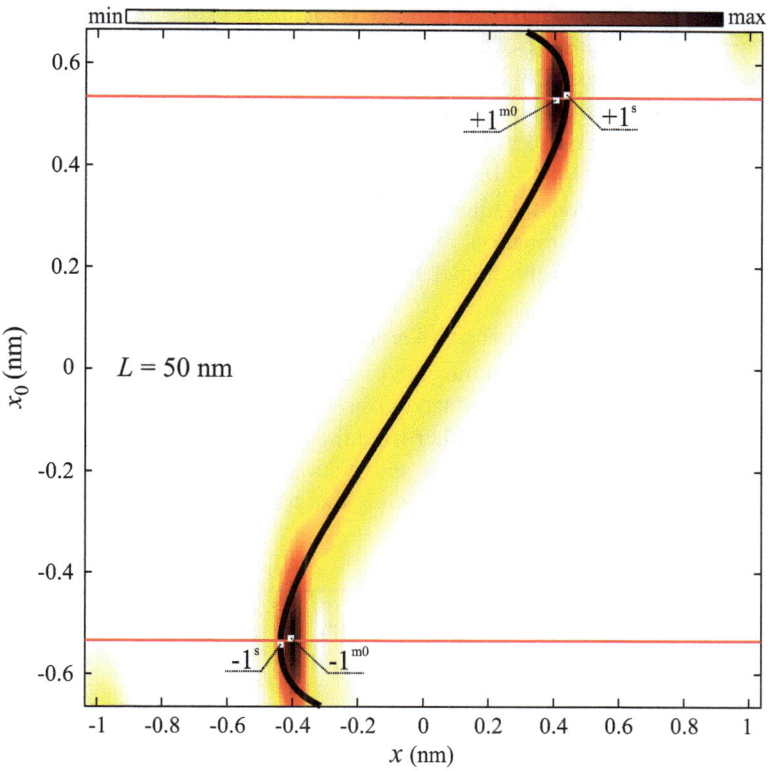


**Fig. 6.5** Two-dimensional representation of the amplitudes squared of the positron wave functions in the final TP plane along the $x$ axis for $-R_n < x_0 < R_n$ and $y_0 = 0$ in the case of 1 MeV positrons and a (11, 9) chiral single-wall carbon nanotube for $L$ = 50 nm, and corresponding positron spatial transmission function [221]. Points $\pm 1^s$ belong to the classical SPR, while $\pm 1^{m0}$ designate the maxima of the quantum mechanical distribution. The regions beyond the red lines correspond to the spatial rainbow subensemble

distance from its wall than the neighboring ray can exit the nanotube at a point very close to the final point of the neighboring ray. Hence, a few focal points appear on the $x$ axis, where the neighboring rays of the wave packet come very close to each other (in the TP plane). The effect is analogous to the one described by Born and Wolf [226]. The positions of the observed maxima or shoulders coincide with the positions of these internal focal points.

It should be emphasized that this explanation of the effect of wave wrinkling is more general than the one given by Petrović et al. [50]. It applies for all the values of $x_0$ for which the effect occurs, while the previous explanation is restricted to the ones close to $x_0^{m0}$.

The positions of the maxima or shoulders of one of the functions $A_\rho^2(x, y = 0, t_f)$ are close to the positions of the corresponding maxima or shoulders of the other functions. Hence, one can say that the wave packets wrinkle in a mutually coordinated way. We explain this by the fact that all the projectiles interact with the same target. For $x_0$ close to $x_0^{m0}$, each function contains a pronounced maximum close to $x^{m0}$. The corresponding wave packet is wrinkled as if a barrier exists at $x^{m0}$, preventing its spreading after the barrier, and, hence, causing its concentration just before the barrier. It is also evident that the curve connecting the positions of the absolute maxima of $A_\rho^2(x, y = 0, t_f)$ is close to the curve representing the function $x(x_0, t_f)$. Therefore, we conclude that the above-described effects of wave wrinkling, concentration, and coordination make the quantum SPR effect. The observed virtual barrier represents the boundary between the bright and dark sides of the rainbow.

Figure 6.6 gives the one-dimensional representation of the functions $A_\rho^2(x, y = 0, t_f)$ for $-R_n < x_0 < R_n$ and $y_0 = 0$ in the case of $L = 50$ nm. In order to make the figure clearer, the number of curves is reduced from $N_p = 301$ to 61. One can see that each of these curves that is centered farther from the origin (for $x_0$ beyond ±0.26 nm) is wrinkled. The curves designated by red color correspond to the wave packets belonging to a spatial rainbow subensemble that will be specified later.

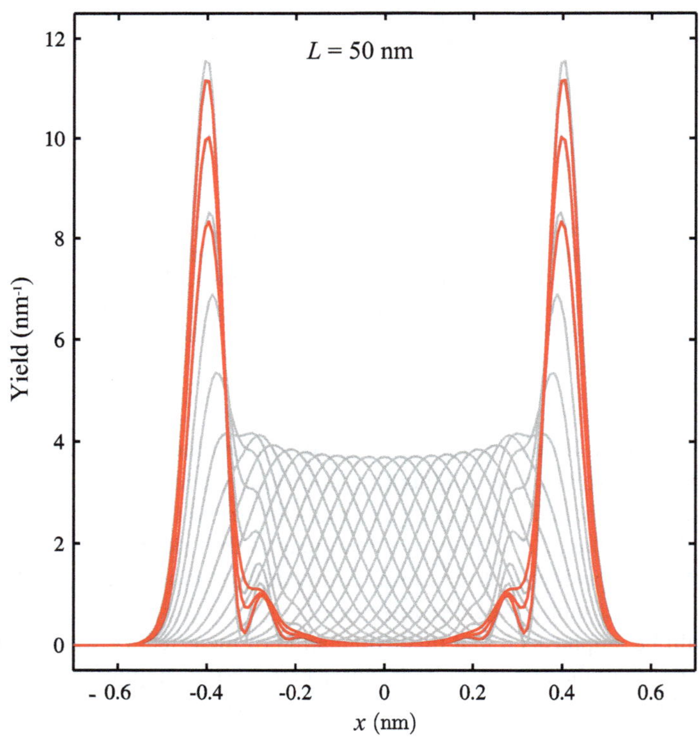


**Fig. 6.6** One-dimensional representation of the amplitudes squared of the positron wave functions in the final TP plane along the $x$ axis for $-R_n < x_0 < R_n$ and $y_0 = 0$ in the case of 1 MeV positrons and a (11, 9) chiral single-wall carbon nanotube for $L = 50$ nm [221]. The *curves* designated by *red color* correspond to the spatial rainbow subensemble

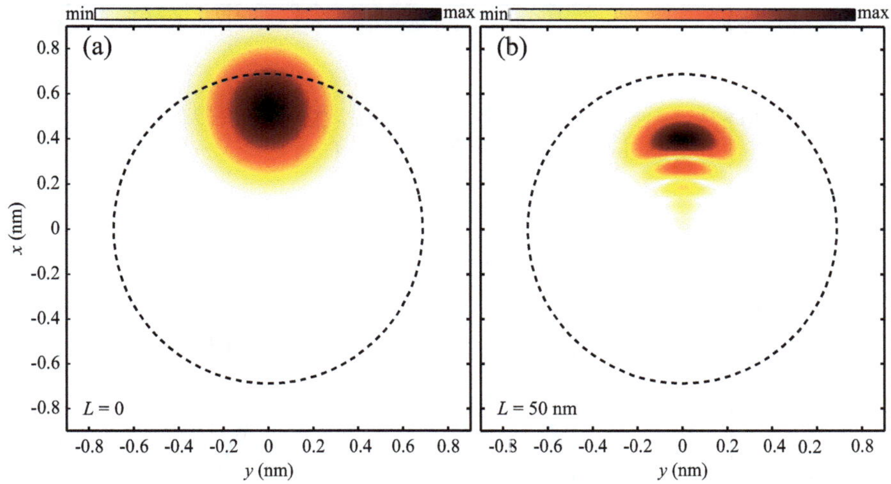


**Fig. 6.7** (**a**) Two-dimensional initial spatial distribution of a 1 MeV positron described as a Gaussian wave packet placed at point (+0.53 nm, 0), and (**b**) resulting spatial distribution of the positron transmitted through a (11, 9) chiral single-wall carbon nanotube for $L = 50$ nm graphed on a logarithmic scale [221]

Figure 6.7a gives the two-dimensional incident spatial distribution of a positron assumed to be a Gaussian wave packet placed at point (+0.53 nm, 0), where the distribution $A_\rho^2(x, y = 0, t_f)$ is maximal. The resulting spatial distribution of the wave packet transmitted through the nanotube of a length of $L = 50$ nm is given in Fig. 6.7b. It contains a strong maximum and a number of weaker maxima extending toward the nanotube axis – it is wrinkled.

Figure 6.8 shows the classical spatial Hamilton principal function of the positrons transmitted through the nanotube divided by $\hbar$, $S_\rho^{\mathrm{cl}}(x, t_f)/\hbar$, in the case of $L = 50$ nm. The curve representing this function has a pair of cusp singular points, $\pm 1^s$, at $x_1^s = \pm 0.44$ nm, corresponding to the maximum and minimum of the function $x(x_0, t_f)$, respectively. It has three branches, joining at the cusp singular points, which will be referred to as its horizontal branch, and its right and left vertical branches.

Figure 6.8 also gives the phases of the positron wave functions in the final TP plane along the $x$ axis, $S_\rho^{\mathrm{qu}}(x, y = 0, t_f)/\hbar$, for $-R_n < x_0 < R_n$ and $y_0 = 0$ in the case of $L = 50$ nm. The number of curves is 61, instead of $N_p = 301$. It is evident that there are several groups of curves representing these functions. The curves from each of these groups extend together with the remaining curves in the directions from the right nanotube wall, at $+R_n = +0.69$ nm, to the right cusp singular point, $+1^s$, and from the left wall, at $-R_n = -0.69$ mm, to the left singular point, $-1^s$, and then separate from the remaining curves and continue "parallel" to the right and left vertical branches of the curve representing the function $S_\rho^{\mathrm{cl}}(x, t_f)/\hbar$, respectively. Let us focus on the curves from the first group, which are most numerous and separate

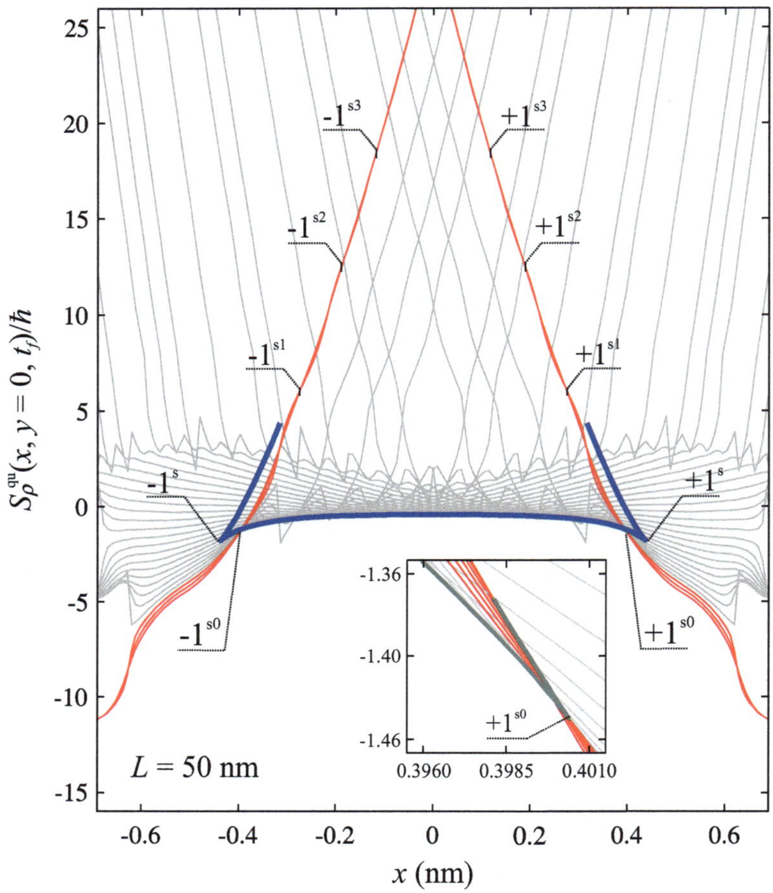


**Fig. 6.8** Phases of the positron wave functions in the final TP plane along the $x$ axis for $-R_n < x_0 < R_n$ and $y_0 = 0$ in the case of 1 MeV positrons and a (11, 9) chiral single-wall carbon nanotube for $L = 50$ nm (*gray* and *red curves*), and corresponding classical spatial Hamilton principal function divided by $\hbar$ (*blue curve*) [221]. The *curves* designated by *red color* correspond to the spatial rainbow subensemble. Points $\pm 1^s$ are the cusp singular points of the classical curve. Points $\pm 1^{s0}$ are the cusp singular points of the envelopes of the quantum mechanical curves in the regions around $\pm 1^s$, and $\pm 1^{s1}$, $\pm 1^{s2}$, and $\pm 1^{s3}$ are the common inflection points of these curves that correspond to the rainbow subensemble. *Inset*: the envelope of the quantum mechanical curves in the region around $+1^s$ (*green curve*), with $+1^{s0}$ as its singular point

first. They are designated by red color. The percentage of curves from this group is 19.3%, and the corresponding values of $x_0$ are beyond $x_0^{r0} = \pm 0.54$ nm going from the origin (see Fig. 6.5). There are 21 additional groups of curves. The percentages of curves from these groups are close to each other as well as the corresponding intervals of values of $x_0$, extending from $x_0^{r0}$ toward the origin. The average percentage of curves from these groups is 3.8%, while the corresponding average interval of values of $x_0$ is 0.03 nm. If the number of shown curves were not reduced, the figure would contain more curves from these groups.

The curves from the right or left subgroup of the first group are close to each other, meaning that the corresponding wave packets are close to being in phase with each other. This means that the behavior of these wave packets is additionally coordinated. Besides, these curves are close to the right or left vertical branch. A detailed analysis of all the curves in the regions around the former cusp singular points, $\pm 1^s$, has shown that their envelope also has a pair of cusp singular points, $\pm 1^{s0}$, at $x_1^{s0} = \pm 0.40$ nm, which are close to $x^{m0}$ (see Fig. 6.5). The "vertical" branches of the right and left parts of the envelope are defined by the curves from the right and left subgroups of the first group, respectively, and its "horizontal" branch by the remaining curves. This is shown in the inset of Fig. 6.8 for the singular point $+1^{s0}$. Thus, the latter singular points are the points of separation of the curves from the first group from the remaining curves. A careful inspection of the curves from the right and left subgroups of the first group after the separation points has shown that each of them has three inflection points, and that the positions of these points coincide with the positions of the corresponding inflection points of the other curves from the subgroup. The positions of these three pairs of common inflection points are $x_1^{s1} = \pm 0.27$ nm, $x_1^{s2} = \pm 0.18$ nm, and $x_1^{s3} = \pm 0.10$ nm.

The same analyses have been performed for the wave packets with $\rho_0 < R_n$ and the other values of $\phi_0$, different from 0 and $\pi$. Its conclusions coincide with those drawn from the analysis for $\phi_0 = 0$ and $\pi$, with the values of $x_1^{s0}$, $x_1^{s1}$, $x_1^{s2}$, and $x_1^{s3}$ coinciding or being close to the ones obtained for $\phi_0 = 0$ and $\pi$.

Now, we can come back to the one-dimensional normalized spatial distribution of transmitted positrons for $L = 50$ nm, given in Fig. 6.4. The positions of the strong maxima coincide with $\pm x_1^{s0}$, and the positions of the weaker maxima are close to $x_1^{s1}$, $x_1^{s2}$, and $x_1^{s3}$. They are denoted as $\pm 1^{s0}$, $\pm 1^{s1}$, $\pm 1^{s2}$, and $\pm 1^{s3}$, respectively. The small differences between the positions of the weaker maxima and $x_1^{s1}$, $x_1^{s2}$, and $x_1^{s3}$, respectively, are attributed to the fact that the distribution is generated with all the values of $x_0$ and $y_0$, rather than with only the values of $x_0$ beyond $x_0^{r0}$ and $y_0 = 0$. The strong maxima belong to a principal SPR and the weaker maxima to three supernumerary SPRs. The positions of the strong maxima are the common positions of the points of separation of all the wave packets from the first group. The positions of the weaker maxima are determined by the common positions of the inflection points of the phases of all the wave packets from the first group. The subensemble comprising the wave packets from the first group has been named the spatial rainbow subensemble.

#### 6.2.2.2 Quantum Spatial Primary Rainbows for $L = 100$ nm

Figure 6.9 shows the one-dimensional normalized spatial distribution of transmitted positrons for $L = 100$ nm. It contains a pair of weaker maxima at $\pm 0.23$ nm and a stronger maximum at the origin. The explanation of this distribution is the same as the one already given for the spatial distribution obtained for $L = 50$ nm.

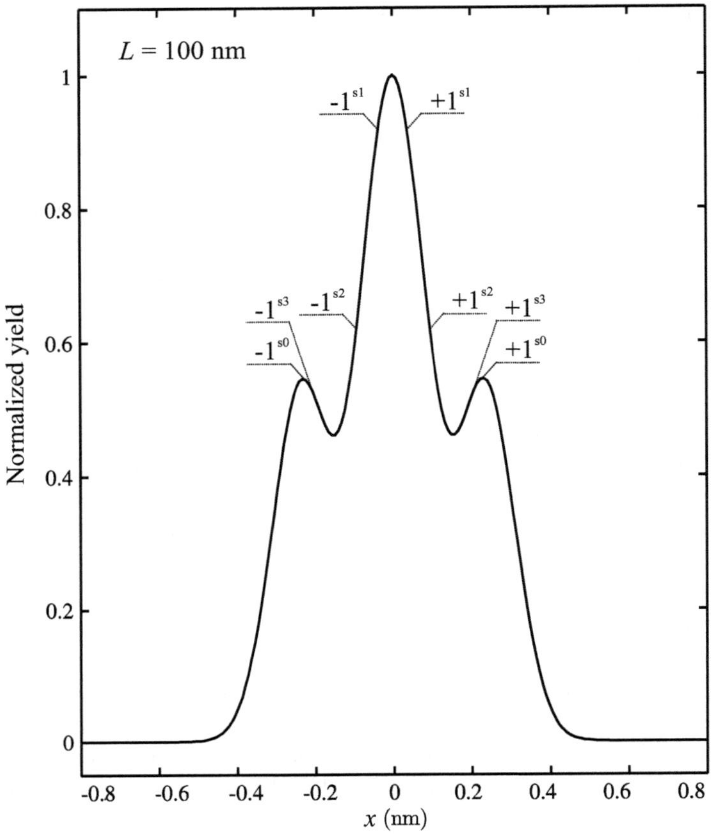


**Fig. 6.9** One-dimensional normalized spatial distribution of 1 MeV positrons transmitted through a (11, 9) chiral single-wall carbon nanotube for $L = 100$ nm represented initially as an ensemble of Gaussian wave packets [221]. The weaker maxima are due to the principal SPR combined with the third supernumerary SPR, $\pm 1^{s0}$ and $\pm 1^{s3}$, and the stronger maximum is due to the first and second supernumerary SPRs, $\pm 1^{s1}$ and $\pm 1^{s2}$, respectively

The function $x(x_0, t_f)$ in the case of $L = 100$ nm is given in Fig. 6.10. It has a maximum and minimum, $\pm 1^s$, at $x_{01}^s = \pm 0.42$ nm, which are $x_1^s = \pm 0.31$ nm, respectively, belonging to the classical SPR. This figure also contains the functions $A_\rho^2(x, y = 0, t_f)$ for $-R_n < x_0 < R_n$ and $y_0 = 0$ in the case of $L = 100$ nm. This two-dimensional distribution has a pair of maxima, $\pm 1^{m0}$, at $x_0^{m0} = \pm 0.42$ nm and $x^{m0} = \pm 0.26$ nm. Analysis has shown that the above-described effects of wave wrinkling, concentration, and coordination appear in this case as well, and that they make the quantum SPR effect. The observed virtual barrier, at $x^{m0}$, is the boundary between the bright and dark sides of the rainbow.

The function $S_\rho^{\mathrm{cl}}(x, t_f)/\hbar$ in the case of $L = 100$ nm is shown in Fig. 6.11. The curve representing this function has a pair of cusp singular points, $\pm 1^s$, at

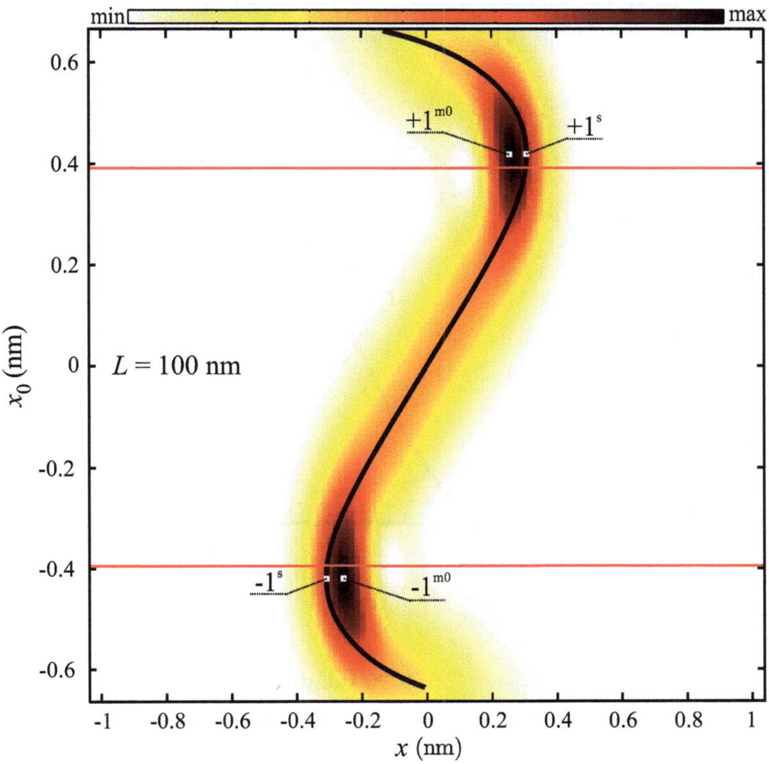


**Fig. 6.10** Two-dimensional representation of the amplitudes squared of the positron wave functions in the final TP plane along the $x$ axis for $-R_n < x_0 < R_n$ and $y_0 = 0$ in the case of 1 MeV positrons and a (11, 9) chiral single-wall carbon nanotube for $L$ = 100 nm, and corresponding positron spatial transmission function [221]. Points $\pm 1^s$ belong to the classical SPR while $\pm 1^{m0}$ designate the maxima of the quantum mechanical distribution. The regions beyond the *red lines* correspond to the spatial rainbow subensemble

$x_1^s = \pm 0.31$ nm, and one crunode singular point, at the origin ($\pm 1^{sa}$). The cusp singular points correspond to the maximum and minimum of the function $x(x_0, t_f)$, respectively, and the crunode singular point corresponds to a pair of points where the curve representing the function $\widetilde{\theta}_x(x, t_f)$ intersects the $\widetilde{\theta}_x$ axis off the origin (see Fig. 6.1).

Figure 6.11 also gives the functions $S_\rho^{\mathrm{qu}}(x, y = 0, t_f)/\hbar$ for $-R_n < x_0 < R_n$ and $y_0 = 0$ in the case of $L$ = 100 nm. The number of curves is 61, instead of $N_p = 301$. The curves representing these functions that belong to the first group, i.e., that correspond to the wave packets from the spatial rainbow subensemble, are designated by red color. The values of $x_0$ in question are beyond $x_0^{r0} = \pm 0.39$ nm going from the origin (see Fig. 6.10). The envelope of all the curves in the regions around the former cusp singular points, $\pm 1^s$, also has a pair of cusp singular points, $\pm 1^{s0}$, at $x_1^{s0} = \pm 0.26$ nm, coinciding with $x^{m0}$ (see Fig. 6.10). These are the points of

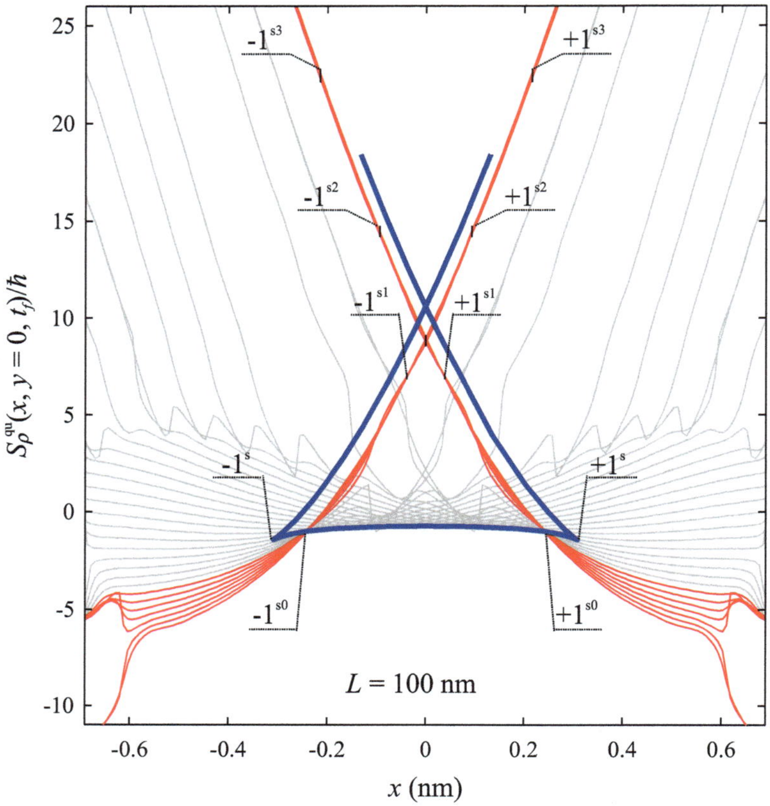


**Fig. 6.11** Phases of the positron wave functions in the final TP plane along the $x$ axis for $-R_n < x_0 < R_n$ and $y_0 = 0$ in the case of 1 MeV positrons and a (11, 9) chiral single-wall carbon nanotube for $L$ = 100 nm (*gray* and *red curves*), and corresponding classical spatial Hamilton principal function divided by $\hbar$ (*blue curve*) [221]. The *curves* designated by *red color* correspond to the spatial rainbow subensemble. Points $\pm 1^s$ are the cusp singular points of the classical curve. Points $\pm 1^{s0}$ are the cusp singular points of the envelopes of the quantum mechanical curves in the regions around $\pm 1^s$, and $\pm 1^{s1}$, $\pm 1^{s2}$, and $\pm 1^{s3}$ are the common inflection points of these curves that correspond to the rainbow subensemble

separation of the curves from the first group from the remaining curves. Each of the curves from the right and left subgroup of the first group has three inflection points after the separation point. The positions of the three pairs of common inflection points of these curves are $x_1^{s1} = \pm 0.04$ nm, $x_1^{s2} = \mp 0.09$ nm, and $x_1^{s3} = \mp 0.22$ nm. The values of $x_1^{s0}$, $x_1^{s1}$, $x_1^{s2}$, and $x_1^{s3}$ obtained for the wave packets with $\rho_0 < R_n$ and the other values of $\phi_0$, different from 0 and $\pi$, coincide or are close to the ones obtained for $\phi_0 = 0$ and $\pi$.

Comparison of the positions of the maxima of the one-dimensional normalized spatial distribution of transmitted positrons for $L$ = 100 nm, given in Fig. 6.9, with $x_1^{s0}$, $x_1^{s1}$, $x_1^{s2}$, and $x_1^{s3}$ leads to the conclusion that the weaker maxima are to be attributed to the principal SPR combined with the third supernumerary SPR, $\pm 1^{s0}$ and $\pm 1^{s3}$, and that the stronger maximum is to be attributed to the first and second supernumerary SPRs, $\pm 1^{s1}$ and $\pm 1^{s2}$, respectively.

#### 6.2.2.3 Quantum Angular Primary Rainbows for $L = 250$ nm

Figure 6.12 shows the one-dimensional normalized angular distribution of transmitted positrons for $L = 250$ nm. One can view it as a superposition of a strong bell-shaped component centered at the origin, and a component consisting of two pairs

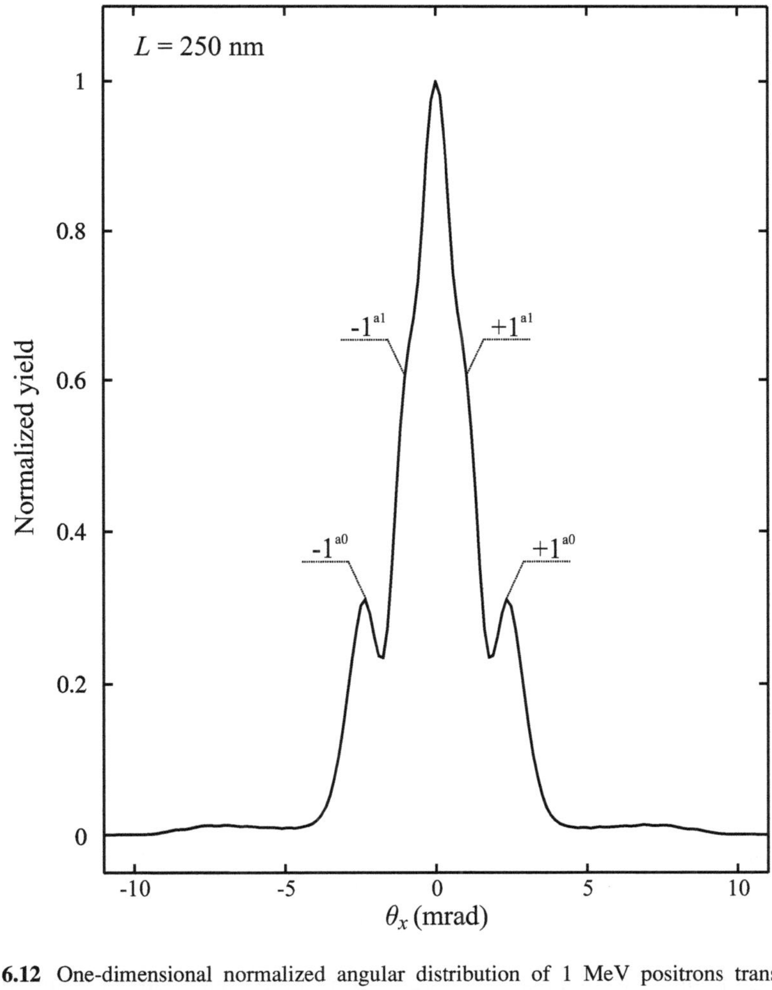


**Fig. 6.12** One-dimensional normalized angular distribution of 1 MeV positrons transmitted through a (11, 9) chiral single-wall carbon nanotube for $L = 250$ nm represented initially as an ensemble of Gaussian wave packets [221]. Maxima $\pm 1^{a0}$ belong to the principal APR, and shoulders $\pm 1^{a1}$ to the first supernumerary APR

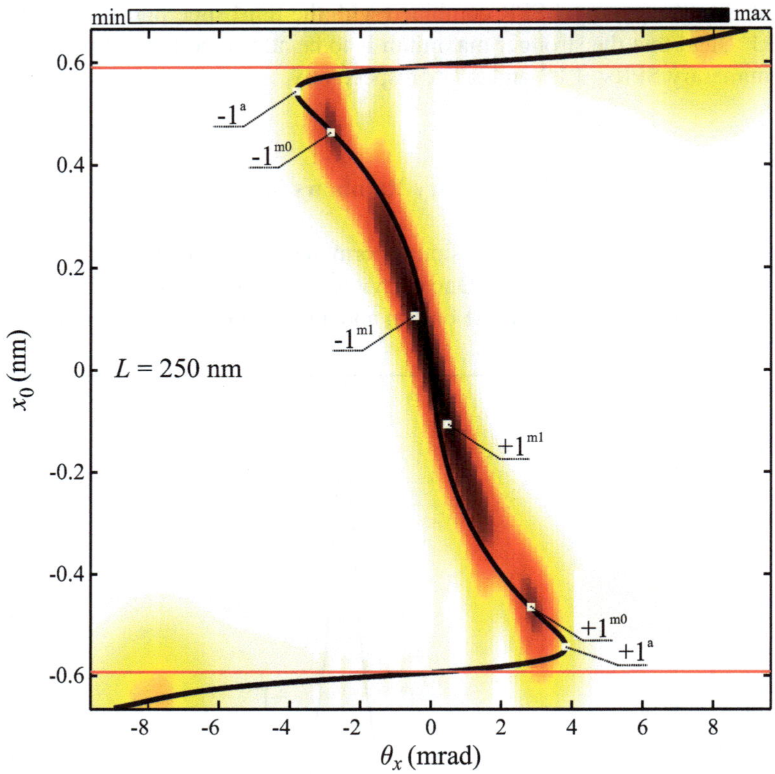


**Fig. 6.13** Two-dimensional representation of the amplitudes squared of the positron wave functions in the TA plane along the $\theta_x$ axis for $-R_n < x_0 < R_n$ and $y_0 = 0$ in the case of 1 MeV positrons and a (11, 9) chiral single-wall carbon nanotube for $L = 250$ nm, and corresponding positron angular transmission function [221]. Points $\pm 1^a$ belong to the classical APR, while $\pm 1^{m0}$ and $\pm 1^{m1}$ designate the maxima of the quantum mechanical distribution. The regions beyond the red lines correspond to the angular rainbow subensemble

of maxima at ±2.3 mrad and ±1.0 mrad, with the second pair of maxima seen in the distribution as a pair of shoulders. We shall explain this distribution in a way analogous to the one already given for the spatial distributions obtained for $L = 50$ and 100 nm.

The positron angular transmission function, $\theta_x(x_0, t_f)$, in the case of $L = 250$ nm is given in Fig. 6.13. It has a maximum and minimum, $\pm 1^a$, at $x^a_{01} = \mp 0.54$ nm, which are $\theta^a_{x1} = \pm 3.8$ mrad, respectively, belonging to the classical APR. This means that for $x_0$ close to $x^a_{01}$, the positrons concentrate in the regions just before $\theta^a_{x1}$ going from the origin. Also, there are no positrons in the regions after $\theta^a_{x1}$. Thus, the region before $\theta^a_{x1}$ is the bright side of the rainbow, and the region after $\theta^a_{x1}$ its dark side. This figure also contains the two-dimensional representation of the amplitudes squared of the positron wave functions in the TA plane along the $\theta_x$ axis, $A^2_\theta(\theta_x, \theta_y = 0, t_f)$, for $-R_n < x_0 < R_n$ and $y_0 = 0$ in the case of $L = 250$ nm. This two-dimensional distribution has two pairs of maxima, $\pm1^{m0}$ and $\pm1^{m1}$, at $x_0^{m0} = \mp0.46$ nm and $\theta_x^{m0} = \pm2.8$ mrad, and at $x_0^{m1} = \mp0.11$ nm and $\theta_x^{m1} = \pm0.5$ mrad, respectively. For $x_0$ close to zero, these functions have the Gaussian shape with the width smaller but close to the initial width. However, for $x_0$ beyond $\pm0.33$ nm going from the origin, each of these functions has a few clearly observable maxima or shoulders – it is wrinkled. This is to be attributed to the effect of internal focusing of the corresponding wave packet. For example, a ray within the wave packet entering the nanotube at a larger angle relative to its wall than the neighboring ray can exit the nanotube at an angle very close to the final angle of the neighboring ray. Hence, a few focal points appear on the $\theta_x$ axis, where the neighboring rays of the wave packet come very close to each other (in the TA plane) [226]. The positions of the observed maxima or shoulders coincide with the positions of these internal focal points.

The positions of the maxima or shoulders of one of the functions $A_\theta^2(\theta_x, \theta_y = 0, t_f)$ are close to the positions of the corresponding maxima or shoulders of the other functions. This means that the wave packets wrinkle in a mutually coordinated way. For $x_0$ close to $x_0^{m0}$, each function contains a pronounced maximum at $\theta_x^{m0}$. The corresponding wave packet is wrinkled as if a barrier exists at $\theta_x^{m0}$, causing its concentration just before the barrier. One can also see that the curve connecting the positions of the absolute maxima of $A_\theta^2(\theta_x, \theta_y = 0, t_f)$ is close to the curve representing the function $\theta_x(x_0, t_f)$. Thus, the conclusion is that the above-described effects of wave wrinkling, concentration, and coordination make the quantum APR effect. The observed virtual barrier represents the boundary between the bright and dark sides of the rainbow.

Figure 6.14a gives the two-dimensional initial angular distribution of a positron assumed to be a Gaussian wave packet placed at point (+0.46 mrad, 0), where the distribution $A_\theta^2(\theta_x, \theta_y = 0, t_f)$ is maximal. The resulting angular distribution of the wave packet transmitted through the nanotube of a length of $L = 250$ nm is given in Fig. 6.14b. It contains a strong maximum and a number of weaker maxima extending toward the origin and beyond – it is wrinkled.

Figure 6.15 shows the classical angular Hamilton principal function of the positrons transmitted through the nanotube divided by $\hbar$, $S_\theta^{\mathrm{cl}}(\theta_x, t_f)/\hbar$, in the case of $L = 250$ nm. The curve representing this function has a pair of cusp singular points, $\pm1^a$, at $\theta_{x1}^a = \pm3.8$ nm, and one crunode singular point, at the origin ($\pm1^{as}$). The cusp singular points correspond to the maximum and minimum of the function $\theta_x(x_0, t_f)$, respectively, and the crunode singular point corresponds to a pair of points where the curve representing the function $\tilde{\theta}_x(x, t_f)$ intersects the $x$ axis off the origin (see Fig. 6.1). The curve representing $S_\theta^{\mathrm{cl}}(\theta_x, t_f)/\hbar$ has three branches, joining at the cusp singular points, which will be referred to as its horizontal branch, and its right and left vertical branches.

Figure 6.15 also gives the phases of the positron wave functions in the TA plane along the $\theta_x$ axis, $S_\theta^{\mathrm{qu}}(\theta_x, \theta_y = 0, t_f)/\hbar$, for $-R_n < x_0 < R_n$ and $y_0 = 0$ in the case of $L = 250$ nm. The number of curves is 61, instead of $N_p = 301$. One can see that there are several groups of curves representing these functions. The curves from each of

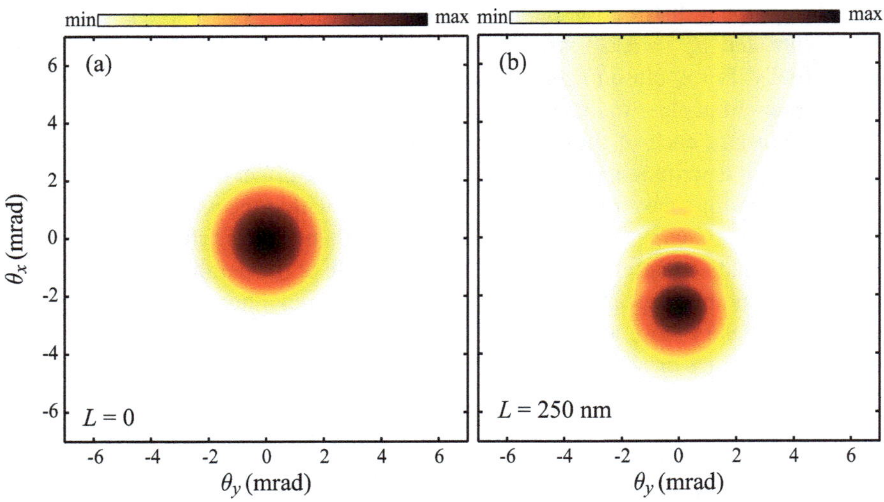


**Fig. 6.14** (**a**) Two-dimensional initial angular distribution of a 1 MeV positron described as a Gaussian wave packet placed at point (+0.46 nm, 0), and (**b**) resulting angular distribution of the positron transmitted through a (11, 9) chiral single-wall carbon nanotube for $L = 250$ nm graphed on a logarithmic scale [221]

these groups extend together with the remaining curves in the directions from the right limit, determined by $+\psi_c= +7.3$ mrad, to the right cusp singular point, $+1^a$, and from the left limit, determined by $-\psi_c$=–7.3 mrad, to the left singular point, $-1^a$, and then separate from the remaining curves and continue "parallel" to the right and left vertical branches of the curve representing the function $S_\theta^{\mathrm{cl}}(\theta_x, t_f)/\hbar$, respectively. We shall focus on the curves from the first group, which separate first. They are designated by red color. The corresponding values of $x_0$ are beyond $x_0^{r0} = \pm 0.59$ nm going from the origin (see Fig. 6.13). The subensemble comprising the wave packets from the first group is called the angular rainbow subensemble.

The curves from the right or left subgroup of the first group are close to each other, meaning that the corresponding wave packets are close to being in phase with each other. This means that the behavior of these wave packets is additionally coordinated. Besides, these curves are close to the right or left vertical branch. A detailed analysis of all these curves in the regions around the former cusp singular points, $\pm 1^a$, has shown that their envelope also has a pair of cusp singular points, $\pm 1^{a0}$, at $\theta_{x1}^{a0} = \pm 3.00$ nm, which are close to $\theta_x^{m0}$ (see Fig. 6.13). The "vertical" branches of the right and left envelopes are defined by the curves from the right and left subgroups of the first group, respectively, and their "horizontal" branches by the remaining curves. This means that the latter singular points are the points of separation of the curves from the first group from the remaining curves. A careful inspection of the curves from the right and left subgroups of the first group after the separation point shows that each of them has several inflection points, and that the positions of these points coincide with the positions of the corresponding inflection

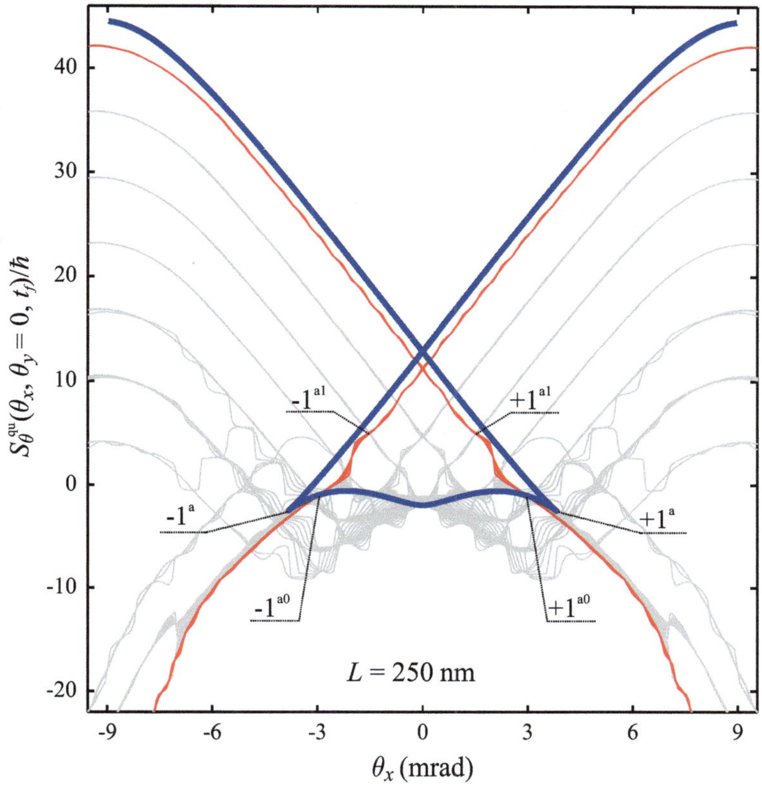


**Fig. 6.15** Phases of the positron wave functions in the TA plane along the $\theta_x$ axis for $-R_n < x_0 < R_n$ and $y_0 = 0$ in the case of 1 MeV positrons and a (11, 9) chiral single-wall carbon nanotube for $L$ = 250 nm (*gray* and *red curves*), and corresponding classical angular Hamilton principal function divided by $\hbar$ (*blue curve*) [221]. The *curves* designated by *red color* correspond to the angular rainbow subensemble. Points $\pm 1^a$ are the cusp singular points of the classical curve. Points $\pm 1^{a0}$ are the cusp singular points of the envelopes of the quantum mechanical curves in the regions around $\pm 1^a$, and $\pm 1^{a1}$ are the common inflection points of these curves that correspond to the rainbow subensemble

points of the other curves from the subgroup. We shall take into account here only the right and left inflection points that lie next to the latter singular points. The positions of the corresponding pair of common inflection points are $\theta_{x1}^{a1} = \pm 1.5$ mrad. The values of $\theta_{x1}^{a0}$ and $\theta_{x1}^{a1}$ obtained for the wave packets with $\rho_0 < R_n$ and the other values of $\phi_0$, different from 0 and $\pi$, coincide or are close to the ones obtained for $\phi_0 = 0$ and $\pi$.

We can now come back to the one-dimensional normalized angular distribution of transmitted positrons for $L$ = 250 nm, given in Fig. 6.12. Its strong bell-shaped component corresponds to the maxima of the distribution $A_\theta^2(\theta_x, \theta_y = 0, t_f)$ at $x_0^{m1}$ and $\theta_x^{m1}$, $\pm 1^{m1}$ (see Fig. 6.13). Since the positions of the maxima and shoulders of the distribution are close to $\theta_{x1}^{a0}$ and $\theta_{x1}^{a1}$, they are denoted as $\pm1^{a0}$ and $\pm1^{a1}$, respectively. The small differences between the positions of the maxima and shoulders and $\theta_{x1}^{a0}$ and $\theta_{x1}^{a1}$, respectively, are attributed to the fact that the distribution is generated with all the values of $x_0$ and $y_0$, rather than with only the values $x_0$ beyond $x_0^{r0}$ and $y_0 = 0$. The maxima belong to the principal APR and the shoulders to a supernumerary APR. The positions of the maxima are the common positions of the points of separation of all the wave packets from the angular rainbow subensemble. The positions of the shoulders are determined by the common positions of the chosen inflection points of the phases of all the wave packets from the angular rainbow subensemble. It has been found that the other inflection points of the phases of the wave packets from the angular rainbow subensemble are connected to the weak supernumerary APRs, and to the principal and supernumerary ASRs, not being well developed. The maxima of the distribution corresponding to the ASRs, lying in the regions after about $\pm5$ mrad going from the origin, can be seen in the figure.

# Epilogue

Here, we shall single out several topics from the material presented in the book that we consider interesting and important for further research and development. Our opinion is that new high-resolution measurements of ion transmission through channels of very thin crystals similar to those conducted by the Singapore group [45–48], complemented with the morphological method introduced in [133], based on the crystal rainbow effect, will lead to additional very accurate ion-atom interaction potentials, which will prove accurate close to the channel axis as well as close to the atomic strings defining the channel. This is explained in Chap. 4.

We have also described in Chap. 4 the effect of superfocusing of channeled ions, which has recently been observed experimentally [134]. Detailed experimental investigations of the possibilities of using the effect for subatomic microscopy are very much needed and can lead to a breakthrough in the area. One can also try to develop the idea of using the superfocusing effect for increasing the luminosity of an ion beam collider.

It has been mentioned in Chap. 5 that a number of theoretical groups have studied ion channeling in carbon nanotubes with the objective to investigate the possibilities of guiding ion beams with nanotubes. We consider experimental studying in this area a real challenge from the points of view of both projectile beam preparation and target fabrication. If successful, such experiments could result in a few advancements in the field of accelerator technologies.

In Chap. 6, we have introduced the quantum mechanical theory of positron rainbow channeling in carbon nanotubes and applied it to three cases of very short nanotubes. Our opinion is that application of the theory in the vicinity of the first superfocusing point is a demanding theoretical task. Ćosić et al. [227] have recently published the first results along this line, related to the cases in which the anharmonicity of the interaction has been very small. In this study, the dependence of the quantum solution of the problem on the anharmonicity and its relation to the classical solution is explored. In addition, a solution of the problem based on catastrophe theory is given and compared with the quantum solution. We prove



N. Nešković et al., *Rainbows in Channeling of Charged Particles in Crystals and Nanotubes*, Lecture Notes in Nanoscale Science and Technology 25,
DOI 10.1007/978-3-319-61524-0

that an effective quantum dynamics of the ensemble can be clearly observed. This theory should be extended to be applied to positron rainbow channeling in planar and axial crystal channels.

It is interesting to note that the theory of crystal rainbows, which has been used extensively in Chaps. 3, 4, 5, and 6 has been also applied in the field of ion optics [228–231]. For example, it has been found that the rainbows also appear in proton transmission through electrostatic and radiofrequency lenses. Such devices can be employed for specific shaping of ion beams. We think that further research and development in this area will definitely give interesting and useful results.

# References


1. N. Nešković (ed.), *Rainbows and Catastrophes*. Proceedings of the Workshop on Rainbow Scattering, Cavtat, 14–16 Aug 1989 (Boris Kidrič Institute of Nuclear Sciences, Belgrade, 1990)
2. H.M. Nussenzveig, The theory of the rainbow. Sci. Am. **236**, 116 (1977)
3. R.G. Newton, *Scattering Theory of Waves and Particles* (Springer, New York, 1982), p. 78
4. H.M. Nussenzveig, High-frequency scattering by a transparent sphere. I. Direct reflection and transmission. J. Math. Phys. **10**, 82 (1969)
5. H.M. Nussenzveig, High-frequency scattering by a transparent sphere. II. Theory of the rainbow and the glory. J. Math. Phys. **10**, 125 (1969)
6. H.M. Nussenzveig, Complex angular momentum theory of the rainbow and the glory. J. Opt. Soc. Am. **69**, 1068 (1979)
7. V. Khare, H.M. Nussenzveig, Theory of the rainbow. Phys. Rev. Lett. **33**, 976 (1974)
8. M.V. Berry, Waves and Thom's theorem. Adv. Phys. **25**, 1 (1976)
9. M. V. Berry, Singularities in waves and rays, Physics of Defects (Les Houches Lectures XXXV), ed. by R. Balian, M. Kléman, J. P. Poirier (North-Holland, Amsterdam, 1981), p. 453
10. J.F. Nye, Optical caustics in the near field from liquid drops. Proc. R. Soc. Lond. A **361**, 21 (1978)
11. J.F. Nye, Optical caustics from liquid drops under gravity: Observations of the parabolic and symbolic umbilics. Philos. Trans. R. Soc. Lond. A **292**, 25 (1979)
12. J.F. Nye, The catastrophe optics of liquid drop lenses. Proc. R. Soc. Lond. A **403**, 1 (1986)
13. R. Thom, *Strutural Stability and Morphogenesis* (Benjamin, Reading, 1975)
14. H.C. Van de Hulst, *Light Scattering by Small Particles* (Wiley, New York, 1957)
15. C.B. Boyer, *The Rainbow: From Myth to Mathematics* (Macmillan, London, 1987)
16. J.D. Jackson, From Alexander of Aphrodisias to Young and Airy. Phys. Rep. **320**, 27 (1999)
17. J.A. Adam, The mathematical physics of rainbows and glories. Phys. Rep. **356**, 229 (2002)
18. J.F. Nye, *Natural Focusing and Fine Structure of Light* (Institute of Physics Publishing, Bristol, 1999)
19. O.B. Firsov, Determination of the forces acting between atoms using the differential effective cross-section for elastic scattering. Zh. Eksp. Teor. Fiz. **24**, 279 (1953) (in Russian)
20. E.A. Mason, Scattering of low velocity molecular beams in gases. J. Chem. Phys. **26**, 667 (1957)
21. K.W. Ford, J.A. Wheeler, Semiclassical description of scattering. Ann. Phys. (New York) **7**, 259 (1959) [Ann. Phys. 281, 608 (2002) (reprinted)]




N. Nešković et al., *Rainbows in Channeling of Charged Particles in Crystals and Nanotubes*, Lecture Notes in Nanoscale Science and Technology 25,
DOI 10.1007/978-3-319-61524-0

22. K.W. Ford, J.A. Wheeler, Application of semiclassical scattering analysis. Ann. Phys. (New York) **7**, 287 (1959)
23. D. Beck, Wide-angle scattering in molecular beams. The rainbow effect. J. Chem. Phys. **37**, 2884 (1962)
24. H.-P. Weise, H.-U. Mittmann, A. Ding, H. Henglein, Streuung von Ionen. II. Regenbogeneffekt bei der elastischen Streuung von Protonen an Helium, Neon, Krypton, Xenon und von He an Helium. Z. Naturforsch. A **26**, 1122 (1971) (in German)
25. N.S. Wall, J.R. Rees, K.W. Ford, Elastic scattering of 22-MeV alpha particles. Phys. Rev. **97**, 726 (1955)
26. H.E. Wegner, R.M. Eisberg, G. Igo, Elastic scattering of 40-MeV alpha particles from heavy elements. Phys. Rev. **99**, 825 (1955)
27. D.A. Goldberg, S.M. Smith, Criteria for the elimination of discrete ambiguities in nuclear optical potentials. Phys. Rev. Lett. **29**, 500 (1972)
28. N. Alamanos, F. Auger, J. Barrette, B. Berthier, B. Fernandez, J. Gastebois, L. Papineau, H. Doubre, W. Mittig, Elastic scattering of $^{40}Ar$ on $^{60}Ni$, $^{120}Sn$ and $^{208}Pb$ at 44 MeV/u. Phys. Lett. B **137**, 37 (1984)
29. H.J. Gils, E. Friedman, H. Rebel, J. Buschmann, S. Zagromski, H. Kiewe-Nebenius, B. Neumann, R. Pesi, G. Bechtold, Nuclear sizes of Ca40,42,44,48 from elastic scattering of 104 MeV alpha paticles. I. Experimental results and optical potentials. Phys. Rev. C **21**, 1239 (1980)
30. R. da Silveira, C. Leclercq-Willain, On the separation of the nuclear and Coulomb rainbow components from the elastic scattering data. Z. Phys. A **314**, 63 (1983)
31. K.W. McVoy, G.R. Satchler, Nuclear rainbows and heavy-ion scattering. Nucl. Phys. A **417**, 157 (1984)
32. D.T. Khoa, W. von Oertzen, H.G. Bohlen, S. Ohkubo, Nuclear rainbow scattering and nucleus-nucleus potential. J. Phys. G Nucl. Part Phys. **34**, R111 (2007)
33. R.A. Oman, Numerical experiments on scattering of noble gases from single-crystal silver. J. Chem. Phys. **48**, 3919 (1968)
34. J.N. Smith Jr., D.R. O'Keefe, H. Saltsburg, P.L. Palmer, Preferential scattering of Ar from LiF: Correlation with lattice properties. J. Chem. Phys. **50**, 4667 (1969)
35. J.D. McClure, Surface rainbows: A similitude between classical and diffractive scattering of atoms from crystalline surfaces. J. Chem. Phys. **52**, 2712 (1970)
36. J.N. Smith Jr., D.R. O'Keefe, P.L. Palmer, Rare-gas scattering from LiF: Correlation with lattice properties. II. J. Chem. Phys. **52**, 315 (1970)
37. A.W. Kleyn, T.C.M. Horn, Rainbow scattering from solid surfaces. Phys. Rep. **199**, 191 (1991)
38. H. Winter, Collisions of atoms and ions with surfaces under grazing incidence. Phys. Rep. **367**, 387 (2002)
39. A. Schüller, H. Winter, Supernumerary rainbows in the angular distribution of scattered projectiles for grazing collisions of fast atoms with a LiF(001) surface. Phys. Rev. Lett. **100**, 097602 (2008)
40. N. Nešković, The effect of transverse correlations in ion channeling in very thin crystals. Oak Ridge Nat. Lab. Rep. **6004**, 203 (1983)
41. N. Nešković, Rainbow effect in ion channeling. Phys. Rev. B **33**, 6030 (1986)
42. H.F. Krause, S. Datz, P.F. Dittner, J. Gomez del Campo, P.D. Miller, C.D. Moak, N. Nešković, P.L. Pepmiller, Rainbow effect in axial ion channeling. Phys. Rev. B **33**, 6036 (1986)
43. H.F. Krause, S. Datz, P.F. Dittner, J. Gomez del Campo, P.D. Miller, C.D. Moak, N. Nešković, P.L. Pepmiller, Rainbow scattering in axial ion-channeling involving very thin crystals: Experimental and theoretical results. Nucl. Instrum. Methods Phys. Res., Sect. B **13**, 51 (1986)
44. H.F. Krause, J.H. Barrett, S. Datz, P.F. Dittner, N.L. Jones, J. Gomez del Campo, C.R. Vane, Angular distribution of ions axially channeled in a very thin crystal: Experimental and theoretical results. Phys. Rev. A **49**, 283 (1994)

45. Z.Y. Dang, M. Motapothula, Y.S. Ow, T. Venkatesan, M.B.H. Breese, M.A. Rana, A. Osman, Fabrication of large-area ultra-thin single crystal silicon membranes. Appl. Phys. Lett. **99**, 223105 (2011)
46. M. Motapothula, Z.Y. Dang, T. Venkatesan, M.B.H. Breese, M.A. Rana, A. Osman, Axial ion channeling patterns from ultra-thin silicon membranes. Nucl. Inst. Methods Phys. Res. B **283**, 29 (2012)
47. M. Motapothula, Z.Y. Dang, T. Venkatesan, M.B.H. Breese, M.A. Rana, A. Osman, Influence of the narrow {111} planes on axial and planar ion channeling. Phys. Rev. Lett. **108**, 195502 (2012)
48. M. Motapothula, Z.Y. Dang, T. Venkatesan, M.B.H. Breese, A study of ion channeling patterns at minor axes in silicon. Nucl. Inst. Methods Phys. Res. B **330**, 24 (2014)
49. S. Petrović, D. Borka, N. Nešković, Rainbows in transmission of high energy protons through carbon nanotubes. Eur. Phys. J. B. **44**, 41 (2005)
50. S. Petrović, M. Ćosić, N. Nešković, Quantum rainbow channeling of positrons in very short carbon nanotubes. Phys. Rev. A **88**, 012902 (2013)
51. D.S. Gemmell, Channeling and related effects in the motion of charged particles through crystals. Rev. Mod. Phys. **46**, 129 (1974)
52. J. Stark, G. Wendt, Über das Eindringen von Kanalstrahlen in feste Körper. Ann. Phys. **38**, 921 (1912) (in German)
53. J. Stark, Bemerkung über Zerstreuung und Absorption von β-Strahlen und Röntgenstrahlen in Kristallen. Phys. Z. **13**, 973 (1912) (in German)


53. J. Stark, Bemerkung über Zerstreuung und Absorption von β-Strahlen und Röntgenstrahlen in Kristallen. Phys. Z. **13**, 973 (1912) (in German)


54. P. K. Rol, J. M. Fluit, F. P. Viehböck, M. de Jong, *Sputtering of copper monocrystals by bombardment with 20 keV $Ar^{+}$*. Proceedings of the Fourth International Conference on Ionization Phenomena in Gases (North-Holland, Amsterdam, 1960), p. 257
55. O. Almén, G. Bruce, Sputtering experiments in the high energy region. Nucl. Inst. Methods **11**, 279 (1961)
56. J.A. Davies, J. Friesen, J.D. McIntyre, A radiochemical technique for studying range-energy relationships for heavy ions of keV energies in aluminum. Can. J. Chem. **38**, 1526 (1960)
57. J.A. Davies, J.D. McIntyre, R.L. Cushing, M. Lounsbury, The range of alkali metal ions of kiloelectron volt energies in aluminum. Can. J. Chem. **38**, 1535 (1960)
58. M.T. Robinson, Deduction of ion ranges in solids from collection experiments. Appl. Phys. Lett. **1**, 49 (1962)
59. M.T. Robinson, O.S. Oen, The channeling of energetic atoms in crystal lattices. Appl. Phys. Lett. **2**, 30 (1963)
60. M.T. Robinson, O.S. Oen, Computer studies of the slowing down of energetic atoms in crystals. Phys. Rev. **132**, 2385 (1963)
61. C. Lehmann, G. Leibfried, Higher order momentum approximations in classical collision theory. Z. Phys. A: Hadrons Nucl. **172**, 465 (1963)
62. C. Lehmann, G. Leibfried, Long-range channeling effects in irradiated crystals. J. Appl. Phys. **34**, 2821 (1963)
63. R.A. Nelson, M.W. Thompson, The penetration of energetic ions through the open channels in a crystal lattice. Philos. Mag. **8**, 1677 (1963)
64. J. Lindhard, Motion of swift charged particles, as influenced by strings of atoms in crystals. Phys. Lett. **12**, 126 (1964)
65. C. Erginsoy, Anisotropic effects in interactions of energetic charged particles in a crystal lattice. Phys. Rev. Lett. **15**, 360 (1965)
66. J. Lindhard, Influence of crystal lattice on motion of energetic charged particles. K. Dan. Vidensk. Selsk., Mat.-Fys. Medd. **34**(14), 1–64 (1965)
67. J.H. Barrett, Brakedown of the statistical-equilibrium hypothesis in channeling. Phys. Rev. Lett. **31**, 1542 (1973)

68. E. Bonderup, H. Esbensen, J.U. Anderson, H.E. Schiøtt, Calculations on axial dechanneling. Radiat. Eff. **12**, 261 (1972)
69. V.V. Beloshitsky, F.F. Komarov, M.A. Kumakhov, Dechanneling, flux-peaking and energy losses of fast charged particles penetrating through thick crystals. Phys. Rep. **139**, 293 (1986)
70. P. Gombas, *Statistische Behandlung des Atoms, in Handbuch der Physik*, vol 36 (Springer, Berlin, 1956), p. 109. (in German)
71. G. Molière, Theorie der Streuung schneller geladener Teilchen I: Einzelstreuung am abgeschirmten Coulomb-Feld. Z. Naturforsch. A **2**, 133 (1947) (in German)
72. J. Lindhard, V. Nielsen, M. Scharff, Approximation methods in classical scattering by screened Coulomb fields. K. Dan. Vidensk. Selsk., Mat.-Fys. Medd. **36**(10), 1–32 (1968)
73. O.B. Firsov, Calculation of the interaction potential of atoms. Sov. Phys. - JETP **6**, 534 (1958)
74. J.F. Ziegler, J.P. Biersack, U. Littmark, *The Stopping and Range of Ions in Solids* (Pergamon Press, New York, 1985)
75. J.F. Ziegler, J.P. Biersack, M.D. Ziegler, *SRIM – The Stopping and Range of Ions in Matter* (SRIM, Annapolis, 2008)
76. J.F. Ziegler, M.D. Ziegler, J.P. Biersack, SRIM – The Stopping and range of ions in matter (2010). Nucl. Inst. Methods Phys. Res. B **268**, 1818 (2010)
77. N. Bohr, The penetration of atomic particles through matter. K. Dan. Vidensk. Selsk., Mat.-Fys. Medd. **18**(8), 1–144 (1948)
78. B.R. Appleton, C. Erginsoy, H.E. Wegner, W.M. Gibson, Channeling effects in the energy loss of 3–11 MeV protons in silicon and germanium single crystals. Phys. Rev. **161**, 330 (1967)
79. M. Blackman, The specific heat of solids, in *Encyclopedia of Physics*, ed. by S. Flügge (Springer, Berlin, 1955), vol 7, part 1, p. 325
80. S. Petrović, S. Korica, M. Kokkoris, N. Nešković, Angular distributions of ions channeled in the <100> Si crystals. Nucl. Inst. Methods Phys. Res. B **193**, 152 (2002)
81. J.H. Barrett, Monte Carlo channeling calculations. Phys. Rev. B **3**, 1527 (1971)
82. J.H. Barrett, Methods of channeling simulation. Nucl. Inst. Methods Phys. Res. B **44**, 367 (1990)
83. P.J.M. Smulders, D.O. Boerma, Computer simulations of channeling in single crystals. Nucl. Inst. Methods Phys. Res. B **29**, 471 (1987)
84. N. Nešković, B. Perović, D. Ćirić, On classical small angle scattering from many-particle targets. Phys. Lett. A **96**, 183 (1983)
85. N. Nešković, B. Perović, Ion channeling and catastrophe theory. Phys. Rev. Lett. **59**, 308 (1987)
86. N. Nešković, B. Perović, The analysis of the rainbow effect in ion channeling by catastrophe theory. Nucl. Inst. Methods Phys. Res. B **33**, 66 (1988)
87. N. Nešković, *Catastrophic ion channeling, in Rainbows and Catastrophes, ed. by N. Nešković. Proceedings of the Workshop on Rainbow Scattering, Cavtat, 14–16 Aug 1989* (Boris Kidrič Institute of Nuclear Sciences, Belgrade, 1990), p. 71
88. N. Nešković, G. Kapetanović, S. Petrović, B. Perović, The $X_9$ catastrophe as the organizing center of crystal rainbows. Phys. Lett. A **179**, 343 (1993)
89. N. Nešković, S. Petrović, G. Kapetanović, B. Perović, W.N. Lennard, Crystal rainbows in the cases of square very thin crystals with one atomic string per primitive cell. Nucl. Inst. Methods Phys. Res. B **93**, 249 (1994)
90. S. Petrović, L. Miletić, N. Nešković, *The catastrophic model of crystal rainbows in the cases of hexagonal very thin crystals with one atomic string per primitive cell.* Book of Contributed Papers of the 17th Summer School and International Symposium on the Physics of Ionized Gases, Belgrade, 29 Aug–1 Sept 1994 (Institute of Physics, Belgrade, 1994), p. 97
91. L. Miletić, S. Petrović, N. Nešković, Decay of zero-degree focusing of channeled ions. Nucl. Inst. Methods Phys. Res. B **115**, 337 (1996)
92. S. Petrović, L. Miletić, N. Nešković, Angular distributions of ions transmitted through a very thin crystal. Nucl. Inst. Methods Phys. Res. B **115**, 348 (1996)

93. S. Petrović, L. Miletić, N. Nešković, *The catastrophic model of crystal rainbows in the cases of centered rectangular very thin crystals with one atomic string per primitive cell*. Book of Contributed Papers of the 18th Summer School and International Symposium on the Physics of Ionized Gases, Belgrade, 2–6 Sept 1996 (Faculty of Sciences of the University of Novi Sad, Novi Sad, 1996), p. 157
94. L. Miletić, S. Petrović, N. Nešković, *Periodicity of the crystal rainbow effect*. Book of Contributed Papers of the 18th Summer School and International Symposium on the Physics of Ionized Gases, Belgrade, 2–6 Sept 1996 (Faculty of Sciences of the University of Novi Sad, Novi Sad, 1996), p. 161
95. L. Miletić, S. Petrović, N. Nešković, Energy dependence of zero-degree focusing of channeled ions. Radiat. Eff. Defects Solids **141**, 235 (1997)
96. L. Miletić, S. Petrović, and N. Nešković, *Crystal rainbows between the first and second rainbow cycles*. Book of Contributed Papers of the 19th Summer School and International Symposium on the Physics of Ionized Gases, Zlatibor, 31 Aug–4 Sept 1998 (Faculty of Physics of the University of Belgrade, Belgrade, 1999), p. 227
97. S. Petrović, L. Miletić, N. Nešković, Theory of rainbows in thin crystals: The explanation of ion channeling applied to $Ne^{10+}$ ions transmitted through a <100> Si thin crystal. Phys. Rev. B **61**, 184 (2000)
98. N. Nešković, S. Petrović, L. Živković, Rainbows with a <111> Si thin crystal. Eur. Phys. J. B. **18**, 553 (2000)
99. S. Korica, S. Petrović, S. Kossionides, N. Nešković, *Periodicity of angular distributions of channeled ions*. Book of Contributed Papers of the 20th Summer School and International Symposium on the Physics of Ionized Gases, Zlatibor, 4–8 Sept 2000 (Institute of Physics, Belgrade, 2000), p. 187
100. L. Živković, S. Petrović, S. Kossionides, N. Nešković, Zero-degree focusing of channeled ions and crystal rainbows. Phys. Lett. A **286**, 292 (2001)
101. N. Nešković, S. Petrović, Crystal rainbows. J. Electron Spectrosc. **129**, 233 (2003)
102. N. Nešković, S. Petrović, Rainbows with crystals and nanotubes. Nanosci. Nanotechnol. Lett. **4**, 1033 (2012)
103. I.V. Komarov, A.P. Shcherbakov, Classical scattering by anisotropic potential. Vestn. Leningr. Gos. Univ. **16**, 24 (1979) (in Russian)
104. Y.N. Demkov, The classical problem of conformal small-angle scattering. Sov. Phys. - JETP **53**, 63 (1981)
105. I. Terzić, N. Nešković, D. Ćirić, Small angle scattering of low energy $K^+$ ions from the polycrystalline Mg and Cu surfaces. Surf. Sci. **88**, L71 (1979)
106. I. Terzić, D. Ćirić, N. Nešković, V.S. Chernysh, B. Perović, Small angle scattering of low energy $K^+$ ions from the (001) Ni surface. Nucl. Inst. Methods **170**, 509 (1980)
107. M.E. Straumanis, E.Z. Aka, Lattice parameters, coefficients of thermal expansion, and atomic weights of purest silicon and germanium. J. Appl. Phys. **23**, 330 (1952)
108. B.W. Batterman, D.R. Chipman, Vibrational amplitudes in germanium and silicon. Phys. Rev. **127**, 690 (1962)
109. Y.S. Umanskii, V.I. Prilepskii, Elastic vibration spectra and characteristic temperatures of germanium and silicon. Sov. Phys. Solid State **7**, 2399 (1966)
110. J. Lindhard, A. Winther, Stopping power of electron gas and equipartition rule. K. Dan. Vidensk. Selsk., Mat.-Fys. Medd. **34**(4), 1–22 (1964)
111. J. Golovchenko, Channeling in very thin crystals, Ph.D. thesis, Rensselaer Polytechnic Institute, Troy, 1972
112. D.D. Armstrong, W.M. Gibson, H.E. Wegner, Proton channeling studies in thin crystals with a supercollimated beam. Radiat. Eff. **11**, 241 (1971)
113. D.D. Armstrong, W.M. Gibson, A. Goland, J.A. Golovchenko, R.A. Levesque, R.L. Meek, H.E. Wegner, A qualitative description of the transverse motion of axial channeled particles in thin crystals. Radiat. Eff. **12**, 143 (1972)
114. N. Nešković, S. Petrović, and M. Ćosić, Unpublished results (2016)

115. M.V. Berry, Cusped rainbow and incoherence effects in the rippling-mirror model for particle scattering from surfaces. J. Phys. A **8**, 566 (1975)
116. V.I. Arnold, Normal forms of functions close to degenerate critical points, the Weyl groups $A_k$, $D_k$ and $E_k$, and Langrange singularities. Funct. Anal. Appl. **6**, 254 (1972)
117. V.I. Arnold, Critical points of smooth functions and their normal forms. Russ. Math. Surv. **30**, 1 (1975)
118. T. Bröcker, *Differentiable Germs and Catastrophes* (Cambridge University Press, Cambridge, 1975)
119. V.I. Arnold, *Catastrophe Theory* (Springer, Berlin, 1984)
120. T. Poston, I. Stewart, *Catastrophe Theory and Its Applications* (Dover Publications, New York, 1996)
121. J.D.J. Donnay, H.M. Ondik, *Crystal Data Determinative Tables*, vol 2 (US Government Publishing Office, Washington, DC, 1973), p. C–44
122. I.R. Porteous, *Geometric Differentiation: For the Intelligence of Curves and Surfaces* (Cambridge University Press, Cambridge, 1994)
123. C. Upstill, Catastrophe optics and caustic networks, Ph.D. thesis, University of Bristol, Bristol, 1979
124. H.W. Streitwolf, *Group Theory in Solid State Physics* (Macdonald, London, 1971)
125. L. Pauling, P. Pauling, *Chemistry* (Freeman, San Francisco, 1975), p. 113
126. I.M. Torrens, *Interatomic Potentials* (Academic, New York, 1972), p. 225
127. H.H. Anderson, J.F. Ziegler, *The Stopping and Range of Ions in Matter*, vol 3 (Pergamon Press, New York, 1977)
128. W.H. Press, S.A. Teukolsky, W.T. Vetterling, B.P. Flannery, *Numerical Recipes in FORTRAN 77: The Art of Scientific Computing* (Cambridge University Press, Cambridge, 1993)
129. S. Petrović, N. Nešković, V. Berec, M. Ćosić, Superfocusing of channeled protons and subatomic measurement resolution. Phys. Rev. A **85**, 032901 (2012)
130. M. Motapothula, S. Petrović, N. Nešković, Z.Y. Dang, M.B.H. Breese, M.A. Rana, A. Osman, Origin of ringlike angular distributions observed in rainbow channeling in ultrathin crystals. Phys. Rev. B **86**, 205426 (2012)
131. J.W. Bruce, P.J. Giblin, *Curves and Singularities* (Cambridge University Press, Cambridge, 1984)
132. X. Artru, S.P. Fomin, N.F. Shulga, K.A. Ispirian, N.K. Zhevago, Carbon nanotubes and fullerites in high-energy and X-ray physics. Phys. Rep. **412**, 89 (2005)
133. S. Petrović, N. Nešković, M. Ćosić, M. Motapothula, M.B.H. Breese, Proton-silicon interaction potential extracted from high-resolution measurements of crystal rainbows. Nucl. Inst. Methods Phys. Res. B **360**, 23 (2015)
134. M. Motapothula, S. Petrović, N. Nešković, T. Venkatesan, M.B.H. Breese, Experimental evidence of the superfocusing effect for axially channeled MeV protons. Phys. Rev. B **94**, 075415 (2016)
135. S. Petrović, N. Starčević, M. Ćosić, and N. Nešković, *Rainbow interaction potentials for proton transmission through the <110> and <111> channels of a silicon crystal* (To be published)
136. P. Tiwald, A. Schüller, H. Winter, K. Tőkési, F. Aigner, S. Gräfe, C. Lemell, J. Burgdörfer, Interaction potentials for fast atoms in front of Al surfaces probed by rainbow scattering. Phys. Rev. B **82**, 125453 (2010)
137. U. Specht, M. Busch, J. Seifert, A. Schüller, H. Winter, K. Gärtner, R. Włodarczyk, M. Sierka, J. Sauer, Rainbow scattering under axial surface channeling from a KCl(001) surface. Phys. Rev. B **84**, 125440 (2011)
138. M. Ćosić, S. Petrović, and N. Nešković, *The inverse proton transmission through a very thin silicon crystal* (To be published)
139. R.B. Alexander, G. Dearnaley, D.V. Morgan, J.M. Poate, The interpretation of channeled-ion beam measurements for foreign atom location in crystals. Phys. Lett. A **32**, 365 (1970)

140. J.U. Andersen, O. Andreasen, J.A. Davies, E. Uggerhøj, The use of channeling-effect techniques to locate interstitial foreign atoms in silicon. Radiat. Eff. **7**, 25 (1971)
141. D. Van Vliet, On the spatial distribution of channelled ions. Radiat. Eff. **10**, 137 (1971)
142. M.A. Kumakhov, A theory of flux peaking effect in channeling. Radiat. Eff. **15**, 85 (1972)
143. R.B. Alexander, P.T. Callaghan, J.M. Poate, Use of the channeling technique and calculated angular distributions to locate Br implanted into Fe single crystals. Phys. Rev. B **9**, 3022 (1974)
144. F. Abel, G. Amsel, M. Bruneaux, C. Cohen, A. L'Hoir, Backscattering study and theoretical investigation of planar channeling processes. I. Experimental results. Phys. Rev. B **12**, 4617 (1975)
145. J.A. Ellison, Flux distribution calculations in planar channeling. Phys. Rev. B **12**, 4771 (1975)
146. F. Abel, G. Amsel, M. Bruneaux, C. Cohen, A. L'Hoir, Backscattering study and theoretical investigation of planar-channeling processes. I. The unperturbed-oscillator model. Phys. Rev. B **13**, 993 (1976)
147. J.A. Ellison, S.T. Picraux, Planar-channeling spatial density under statistical equilibrium. Phys. Rev. B **18**, 1028 (1978)
148. M.A. Kumakhov, G. Shirmer, *Atomic Collisions in Crystals* (Atomizdat, Moscow, 1980) (in Russian) [(Gordon and Breach, New York, 1989)]
149. L.C. Feldman, MeV ion scattering for surface structure determination. Crit. Rev. Solid State Mater. Sci. **10**, 143 (1981)
150. L.C. Feldman, J.W. Mayer, S.T. Picraux, *Materials Analysis by Ion Channeling* (Academic Press, New York, 1982)
151. G.V. Dedkov, A.M. Kumakhov, M.Z. Sokhov, The study of the flux-peaking effect of the planar channeling of protons in silicon by computer simulation. Radiat. Eff. **71**, 261 (1983)
152. Y.N. Demkov, J.D. Meyer, A sub-atomic microscope, superfocusing in channeling and close encounter atomic and nuclear reactions. Eur. Phys. J. B. **42**, 361 (2004)
153. N. Nešković, S. Petrović, D. Borka, Superfocusing of channeled protons and crystal rainbows. Nucl. Inst. Methods Phys. Res. B **267**, 2616 (2009)
154. D. Borka, S. Petrović, V. Lukić, J. Timko, V. Berec, N. Nešković, *Channeling of protons in a <100> Si thin crystal: A quantum mechanical approach*. Proceedings of the 25th Summer School and International Symposium on the Physics of Ionized Gases (Publication 89), Donji Milanovac, 30 Aug–3 Sept 2010 (Astronomical Observatory of Belgrade, Belgrade, 2010), p. 89
155. A. Denker, J. Opitz-Coutureau, J.L. Campbell, J.A. Maxwell, T. Hopman, High-energy PIXE: Quantitative analysis. Nucl. Inst. Methods Phys. Res. B **219-220**, 130 (2004)
156. A. Denker, W. Bohne, J.L. Campbell, P. Heide, T. Hopman, J.A. Maxwell, J. Opitz-Coutureau, J. Rauschenberg, J. Röhrich, E. Strub, High-energy PIXE using very energetic protons: Quantitative analysis and cross-sections. X-Ray Spectrom. **34**, 376 (2005)
157. J.C. Slater, Atomic radii in crystals. J. Chem. Phys. **41**, 3199 (1964)
158. J. Lahtinen, P. Kantola, S. Jaatinen, K. Habermehl-Cwirzen, P. Salo, J. Vuorinen, M. Lindroos, K. Pussi, A.P. Seitsonen, LEED and DFT investigation on the $(2 \times 2)$-S overlay on Co(0001). Surf. Sci. **599**, 113 (2005)
159. C. Froese Fischer, *The Hartree-Fock Method for Atoms* (Wiley, New York, 1997), p. 33
160. R. Jebasinski, S. Mantl, C. Dieker, W. Jäger, Ion scattering investigations of buried $CoSi_2$ layers produced by ion beam synthesis. Nucl. Inst. Methods Phys. Res. B **64**, 99 (1992)
161. H. Nolte, W. Assmann, H. Huber, S.A. Karamian, H.D. Mieskes, Blocking- and channeling-ERDA with heavy ions. Nucl. Inst. Methods Phys. Res. B **136**, 587 (1998)
162. M.B.H. Breese, D.N. Jamieson, P.J.C. King, *Materials Analysis Using a Nuclear Microprobe* (Wiley, New York, 1996), p. 59
163. D. Borka, S. Petrović, S. Kossionides, N. Nešković, *Doughnut effect with a hexagonal thin crystal*. Book of Contributed Papers of the 20th Summer School and International

Symposium on the Physics of Ionized Gases, Zlatibor, 4–8 Sept 2000 (Institute of Physics, Belgrade, 2000), p. 183
164. N. Nešković, S. Petrović, D. Borka, S. Kossionides, Rainbows with a tilted <111> Si very thin crystal. Phys. Lett. A **304**, 114 (2002)
165. D. Borka, S. Petrović, S. Kossionides, N. Nešković, *Transmission patterns of protons channeled through a tilted hexagonal thin crystal*. Book of Contributed Papers of the 21th Summer School and International Symposium on the Physics of Ionized Gases, Sokobanja, 26–30 Aug 2002 (Faculty of Sciences and Mathematics of the University of Niš, Niš, 2002), p. 210
166. L.T. Chadderton, Diffraction and channeling. J. Appl. Crystallogr. **3**, 429 (1970)
167. J.S. Rosner, W.M. Gibson, J.A. Golovchenko, A.N. Goland, H.E. Wegner, Quantitative study of the transmission of axially channeled protons in thin silicon crystals. Phys. Rev. B **18**, 1066 (1978)
168. S.K. Andersen, O. Fich, H. Nielsen, H.E. Schiøtt, E. Uggerhøj, C. Vraast Thomsen, G. Charpak, G. Petersen, F. Sauli, J.P. Ponpon, P. Siffert, Influence of channeling on scattering of 2-15 GeV/c protons, $\pi^+$, and $\pi^-$ incident on Si and Ge crystals. Nucl. Phys. B **167**, 1 (1980)
169. D. Borka, S. Petrović, N. Nešković, Doughnuts with a <110> very thin Si crystal. J. Electron Spectrosc. **129**, 183 (2003)
170. S. Petrović, M. Ćosić, N. Nešković, *Classical and quantum rainbow channeling of charged particles in very thin crystals and carbon nanotubes*. Proceedings of the XXII International Conference on Ion-Surface Interactions (ISI 2015), Moscow, 20–24 Aug 2015 (National Research Nuclear University MEPhI, Moscow, 2015), p. 153
171. S. Iijima, Helical microtubules in graphitic carbon. Nature **354**, 56 (1991)
172. R. Saito, G. Dresselhaus, M.S. Dresselhaus, *Physical Properties of Carbon Nanotubes* (Imperial College Press, London, 2001)
173. R.H. Baughman, A.A. Zakhidov, W.A. de Heer, Carbon nanotubes – The route toward applications. Science **297**, 787 (2002)
174. V.V. Klimov, V.S. Letokhov, Hard X-radiation emitted by a charged particle moving in a carbon nanotube. Phys. Lett. A **222**, 424 (1996)
175. L.A. Gevorgian, K.A. Ispirian, R.K. Ispirian, High energy particle channeling in nanotubes. Nucl. Instrum. Methods Phys. Res., Sect. B **145**, 155 (1998)
176. N.K. Zhevago, V.I. Glebov, Channeling of fast charged and neutral particles in nanotubes. Phys. Lett. A **250**, 360 (1998)
177. V.M. Biryukov, S. Bellucci, Nanotube diameter optimal for channeling of high-energy particle beam. Phys. Lett. B **542**, 111 (2002)
178. V.M. Biryukov, Y.A. Chesnokov, V. Guidi, W. Scandale, Channeling of high energy beams in nanotubes. Nucl. Instrum. Methods Phys. Res., Sect. B **202**, 236 (2003)
179. A.A. Greenenko, N.F. Shulga, Fast ion passing through straight and bent nanotubes. Nucl. Instrum. Methods Phys. Res., Sect. B **205**, 767 (2003)
180. N.K. Zhevago, V.I. Glebov, Computer simulations of fast particle propagation through straight and bent nanotubes. Phys. Lett. A **310**, 301 (2003)
181. D. Borka, S. Petrović, N. Nešković, *Angular distributions of protons transmitted through carbon nanotubes*. Proceedings of the Fifth General Conference of the Balkan Physical Union, Section 6: Condensed Matter Physics, Vrnjačka Banja, 25–29 Aug 2003 (Serbian Physical Society, Belgrade, 2003), p. 625
182. S. Bellucci, V.M. Biryukov, A. Cordelli, Channeling of high-energy particles in a multi-wall nanotube. Phys. Lett. B **608**, 53 (2005)
183. A.V. Krasheninnikov, K. Nordlund, Multiwalled carbon nanotubes as apertures and conduits for energetic ions. Phys. Rev. B **71**, 245408 (2005)
184. C.S. Moura, L. Amaral, Channeling on carbon nanotubes: A molecular dynamics approach. J. Phys. Chem. B **109**, 13515 (2005)
185. N. Nešković, S. Petrović, D. Borka, Angular distributions of 1 GeV protons channeled in bent short single-wall carbon nanotubes. Nucl. Instrum. Methods Phys. Res., Sect. B **230**, 106 (2005)

186. S. Petrović, D. Borka, N. Nešković, Rainbow effect in channeling of high energy protons through single-wall carbon nanotubes. Nucl. Instrum. Methods Phys. Res., Sect. B **234**, 78 (2005)
187. D. Borka, S. Petrović, N. Nešković, Rainbow effect in channeling of high energy protons in (10, 0) single-wall carbon nanotubes. Mater. Sci. Forum **494**, 89 (2005)
188. D. Borka, S. Petrović, N. Nešković, *Axial channeling of high energy protons in carbon nanotubes*. Proceeding of the 1st International Workshop on Nanoscience and Nanotechnology (IWON 2005), Belgrade, 15–18 Nov 2005 (Institute of Chemistry, Technology and Metallurgy, Belgrade, 2005), p. 119
189. D. Borka, S. Petrović, N. Nešković, D.J. Mowbray, Z.L. Mišković, Influence of the dynamical image potential on the rainbows in ion channeling through short carbon nanotubes. Phys. Rev. A **73**, 062902 (2006)
190. D. Borka, S. Petrović, N. Nešković, Channeling star effect with bundles of carbon nanotubes. Phys. Lett. A **354**, 457 (2006)
191. D. Borka, S. Petrović, N. Nešković, D.J. Mowbray, Z.L. Mišković, Influence of the dynamic polarization effect on the angular distributions of protons channeled in double-wall carbon nanotubes. Nucl. Instrum. Methods Phys. Res., Sect. B **256**, 131 (2007)
192. S.I. Matyukhin, K.Y. Frolenkov, Critical parameters of channeling in nanotubes. Tech. Phys. Lett. **33**, 58 (2007)
193. Z.L. Mišković, Ion channeling through carbon nanotubes. Radiat. Eff. Defects Solids **162**, 185 (2007)
194. C.S. Moura, L. Amaral, Carbon nanotubes proposed as particle pipes. Carbon **45**, 1802 (2007)
195. D. Borka, D.J. Mowbray, Z.L. Mišković, S. Petrović, N. Nešković, Dynamic polarization effects on the angular distributions of protons channeled through carbon nanotubes in dielectric media. Phys. Rev. A **77**, 032903 (2008)
196. D. Borka, D.J. Mowbray, Z.L. Mišković, S. Petrović, N. Nešković, Channeling of protons through carbon nanotubes embedded in dielectric media. J. Phys. Condens. Matter **20**, 474212 (2008)
197. D. Borka, D.J. Mowbray, Z.L. Mišković, S. Petrović, N. Nešković, Channeling of protons through carbon nanotubes. J. Phys. Conf. Ser. **133**, 012015 (2008)
198. I. Telečki, S. Petrović, D. Borka, N. Nešković, *Axial channeling of high energy protons in carbon nanotubes*. Proceedings of the 24th Summer School and International Symposium on the Physics of Ionized Gases (Publication 84), Novi Sad, 25–29 Aug 2008 (Astronomical Observatory of Belgrade, Belgrade, 2008), p. 173
199. S.I. Matyukhin, Dechanneling of ions on atoms incorporated into carbon nanotubes. Tech. Phys. Lett. **34**, 835 (2008)
200. S. Petrović, D. Borka, N. Nešković, Rainbows with carbon nanotubes. Adv. Stud. Theor. Phys. **2**, 415 (2008)
201. S. Petrović, I. Telečki, D. Borka, N. Nešković, Proton channeling through long chiral carbon nanotubes: The rainbow route to equilibration. Phys. Lett. A **372**, 6003 (2008)
202. L.-P. Zheng, Z.-Y. Zhu, Y. Li, D.-Z. Zhu, H.-H. Xia, Isotopic mass effects for low-energy ion channeling in single-wall carbon nanotubes. J. Phys. Chem. C **112**, 15204 (2008)
203. L.-P. Zheng, Z.-Y. Zhu, Y. Li, D.-Z. Zhu, H.-H. Xia, Ion mass dependence for low energy channeling in single-wall nanotubes. Nucl. Inst. Methods Phys. Res. B **266**, 849 (2008)
204. S. Petrović, D. Borka, I. Telečki, N. Nešković, Angular distributions of high energy protons channeled in long (10, 10) single-wall carbon nanotubes. Nucl. Inst. Methods Phys. Res. B **267**, 2365 (2009)
205. D. Borka, D.J. Mowbray, Z.L. Mišković, S. Petrović, N. Nešković, Donut and dynamic polarization effects in proton channeling through carbon nanotubes. New J. Phys. **12**, 043021 (2010)
206. N. Nešković, D. Borka, S. Šopić, S. Petrović, Rainbows in channeling of 1 GeV protons in a bent very short (11, 9) single-wall carbon nanotube. Int. J. Nonlin. Sci. Num. Simul. **11**, 1131 (2010)
207. Zh. Zhu, D. Zhu, R. Lu, Zh. Xu, W. Zhang, and H. Xia, *The experimental progress in studying of channeling of charged particles along nanostructures*. Proceedings of the International

Conference on Charged and Neutral Particles Channeling Phenomena (Vol. 5974) (SPIE, Bellingham, Washington, 2005), p. 13
208. G. Chai, H. Heinrich, L. Chow, T. Schenkel, Electron transport through single carbon nanotubes. Appl. Phys. Lett. **91**, 103101 (2007)
209. A.S. Berdinsky, P.S. Alegaonkar, J.B. Yoo, H.C. Lee, J.S. Jung, J.H. Han, D. Fink, L.T. Chadderton, Growth of carbon nanotubes in etched ion tracks in silicon oxide on silicon. Nano **02**, 59 (2007)
210. D. Borka, S. Petrović, N. Nešković, *Channeling of Protons Through Carbon Nanotubes* (Nova Science Publishers, New York, 2011)
211. A. Thess, R. Lee, P. Nikolaev, H. Dai, P. Petit, J. Robert, C. Xu, Y.H. Lee, S.G. Kim, A.G. Rinzler, D.T. Colbert, G. Scuseria, D. Tománek, J.E. Fischer, R.E. Smalley, Crystalline ropes of metallic carbon nanotubes. Science **273**, 483 (1996)
212. J. Hone, B. Batlogg, Z. Benes, A.T. Johnson, J.E. Fischer, Quantized phonon spectrum of single-wall carbon nanotubes. Science **289**, 1730 (2000)
213. L. Rossi, Superconductivity and the LHC: The early days. CERN Courrier **51**(9), 21 (2011)
214. M. Ćosić, S. Petrović, N. Nešković, *Spatial rainbows with protons and very long carbon nanotubes* (To be published)
215. M.V. Berry, D.H.J. O'Dell, Ergodicity in wave-wave diffraction. J. Phys. A Math. Gen. **32**, 3571 (1999)
216. J.U. Andersen, W.M. Augustyniak, E. Uggerhøj, Channeling of positrons. Phys. Rev. B **3**, 705 (1971)
217. M.J. Pedersen, J.U. Anderson, W.M. Augustyniak, Channeling of positrons. Radiat. Eff. **12**, 47 (1972)
218. R. Haakwnaasen, L.V. Hau, J.A. Golovchenko, J.C. Palathingal, J.P. Peng, P. Asoka-Kumar, K.G. Lynn, Quantum channeling effects for 1 MeV positrons. Phys. Rev. Lett. **75**, 1650 (1995)
219. A. Howie, M.J. Whelan, Diffraction contrast of electron microscope images of crystal lattice defects. II. The development of a dynamical theory. Proc. R. Soc. Lond. A **263**, 217 (1961)
220. M. Ćosić, S. Petrović, N. Nešković, Quantum rainbow characterization of short chiral carbon nanotubes. Nucl. Inst. Methods Phys. Res. B **323**, 30 (2014)
221. M. Ćosić, S. Petrović, N. Nešković, Quantum primary rainbows in transmission of positrons through very short carbon nanotubes. Nucl. Inst. Methods Phys. Res. B **373**, 52 (2016)
222. J.A. Ellison, S.T. Picraux, W.R. Allen, W.K. Chu, Planar channeling in superlattices: Theory. Phys. Rev. B **37**, 7290 (1988)
223. H. Goldstein, *Classical Mechanics* (Addison-Wesley, Reading, 1982), p. 386
224. M. Ćosić, S. Petrović, N. Nešković, Computational method for the long time propagation of quantum channeled particles in crystals and carbon nanotubes. Nucl. Inst. Methods Phys. Res. B **330**, 33 (2014)
225. H. Weyl, *The Group Theory and Quantum Mechanics* (Dover Publications, New York, 1950), p. 93
226. M. Born, E. Wolf, *Principles of Optics: Electromagnetic Theory of Propagation, Interference and Diffraction of Light* (Cambridge University Press, Cambridge, 1999), p. 134
227. M. Ćosić, S. Petrović, N. Nešković, Effective quantum dynamics in a weakly anharmonic interaction in the vicinity of a focusing point. Nucl. Inst. Methods Phys. Res. B **399**, 1 (2017)
228. N. Nešković, I. Telečki, B. Bojović, S. Petrović, A square electrostatic rainbow lens: Catastrophic ion beam focusing. Nucl. Inst. Methods Phys. Res. A **635**, 1 (2011)
229. I. Telečki, S. Petrović, P. Beličev, B. Rađenović, R. Balvanović, B. Bojović, N. Nešković, Focusing properties of a square electrostatic rainbow lens. Nucl. Inst. Methods Phys. Res. A **694**, 224 (2012)
230. N. Nešković, P. Beličev, I. Telečki, S. Petrović, Rainbow lenses, in *Advances in Imaging and Electron Physics*, vol. 182, (Academic Press, Elsevier, Waltham, 2014), p. 123
231. I. Telečki, P.D. Beličev, S.M. Petrović, N.B. Nešković, Focusing properties of a square electrostatic rainbow lens doublet. Nucl. Technol Radial. Prot. **30**, 239 (2015)

# Author Index






N. Nešković et al., *Rainbows in Channeling of Charged Particles in Crystals and Nanotubes*, Lecture Notes in Nanoscale Science and Technology 25,
DOI 10.1007/978-3-319-61524-0

# Subject Index





N. Nešković et al., *Rainbows in Channeling of Charged Particles in Crystals and Nanotubes*, Lecture Notes in Nanoscale Science and Technology 25,
DOI 10.1007/978-3-319-61524-0

MIX
Papier aus verantwortungsvollen Quellen
Paper from responsible sources
FSC® C105338

If you have any concerns about our products,
you can contact us on
**ProductSafety@springernature.com**

In case Publisher is established outside the EU,
the EU authorized representative is:
**Springer Nature Customer Service Center GmbH**
**Europaplatz 3, 69115 Heidelberg, Germany**

Printed by Libri Plureos GmbH
in Hamburg, Germany